Elements of Meteorology

Vilhelm F. K. Bjerknes (1862–1951). Dr. Bjerknes was one of the great pioneers who created the science of dynamic meteorology. His first research was concerned with the Hertzian electromagnetic waves and their hydrodynamic analysis. When he transferred his interest to meteorology, he applied sound physical principles to the study of atmospheric flow, advancing rapidly the study of theoretical meteorology. His textbook, Dynamical Meteorology and Hydrology, became an internationally known authority. Then, during World War I, he led the Bergen school of meteorologists into the frontal theory of cyclones, a development that revolutionized weather forecasting.

Elements of Meteorology

RICHMOND W. LONGLEY

The University of Alberta

JOHN WILEY & SONS, INC.

New York · London · Sydney · Toronto

PREFACE

This book has developed through ten years of teaching classes of under-graduates about weather and climate. Meteorology as a science is basi-cally physics and deals with matters such as fluid dynamics, radiation, and thermodynamics. Yet many university students wish to learn about weather but lack any knowledge of university mathematics and physics. It then becomes necessary to select topics and developments that will be meaningful to them without further university training.

The basic tools of the meteorologist are the upper-air diagram and the weather maps (surface and upper air). This book has been planned, first, to develop in the student an understanding and an appreciation of these tools. Second, with these tools as a basis for his understanding, the student observes how they can be used to explain some of the common and less common weather phenomena. Third, the text discusses the flows of energy that arise from the atmospheric circulation and analyses the hypotheses that have been advanced to explain changes in the circulation and with it the climate.

This book has been written with the expectation that the student has an understanding of high school mathematics and physics. Mathematical equations related to physical laws are introduced, explained, and used in the development. Problems are given at the end of the chapters. Some of them are relatively simple, but will be useful in testing the student's comprehension. Other problems will challenge even the better students, but the results should prove enlightening.

In general, the centimeter-gram-second system of units and degrees Celsius are used in the text examples and exercises because of the ease of computation in this system, although the English system is sometimes introduced. In giving derived units, such as grams per cubic centimeter, use is made of negative indexes ($gm\ cm^{-3}$). Because of its clarity and simplicity, this system of terminology is becoming standard in scientific texts. Yet it may be new to some high school graduates and may cause difficulties until they become familiar with it.

In reviewing the work that has gone into the preparation of this book, I am grateful for the help that I have received. I have been assisted by many people, some of whose help I have been unable to use. My thanks go to all of them wherever they may be. I am particularly grateful to the contributors of copyright material. My thanks also go to Mr. J. R. H. Noble, Director, Meteorological Service of Canada and to members of his staff, who provided much material and advice in the preparation of the text and diagrams, and to my colleagues, Dr. E. R. Reinelt and Dr. K. D. Hage, for their wise critical comments. My secretary, Mrs. D. C. Keehn, has been admirably patient with my demands on her skill and training.

University of Alberta   RICHMOND W. LONGLEY
Edmonton, Alberta, Canada

CONTENTS

Elements of Meteorology

CHAPTER 1

WHY THE WINDS BLOW

1.1 *Introduction*

Mankind, early in his development, must have become aware of the state of the atmosphere about him, and the changes that occurred from day to day and from season to season. Long before the dawn of history, he began to look for explanations of the things he observed and, in doing so, gave birth to the science of meteorology. The original explanations have been replaced by others as more knowledge about the weather has been obtained. In this, meteorology is similar to all sciences.

In seeking an explanation, people have observed one event subsequent to another, and considered that the second was a consequence of the first. A frost with a new moon came, they said, because the change in the moon brought a change in the weather. Similarly, abnormal weather during the past decade has been blamed on atomic explosions. Other people have seen hail fall on their enemies, or drought destroying their crops, and have explained them as manifestations of the pleasure or disapproval of the storm god.

In the search for causes, meteorologists have discarded many of the explanations given by those not trained in the science. They are still not able to understand fully many weather phenomena. The basic elements, as they view them, are the sun, the earth, and the gases enveloping the earth. Their relationships are complex, with the result that the progress in understanding weather has been relatively slow. In this chap-

1

ter a general picture of the weather over the earth is presented so that the reader may grasp something of the magnitude and complexity of the problem. It is necessary to introduce some ideas without a full explanation, but in the subsequent chapters these and other topics are examined more carefully and in detail in order to present more clearly the weather phenomena and their causes.

1.2 *The Sun as a Heat Source*

As every junior high school pupil knows, the sun is the primary cause for the weather on the earth. In the total universe, the sun is relatively a small star; it is important to us because it is the only star nearby. The average distance from the earth to the sun is 1.5×10^8 km (9.3×10^7 mi). The sun's diameter is 1.4×10^6 km, and its mass 3.32×10^5 times that of the earth.

The most significant fact about the sun, when one considers the weather and climate of the earth, is its apparent surface temperature (approximately 5600°C). As all hot bodies do, it sends off large amounts of energy in all directions. Enough of this energy is intercepted by the earth and changed into heat to maintain life as we know it on our planet. Outside of the earth's atmosphere, the energy falling on one square centimeter at the earth's average distance from the sun is equivalent to 2 cal every minute, a value called the *solar constant*.

Closer to the sun the flow of heat is much greater, the value at the orbit of Venus being 3.72 cal cm^{-2} min^{-1}, or nearly twice as much per unit area as at the earth. On the planet Mars, the heat flow is only 0.84 cal cm^{-2} min^{-1}, or less than one half the amount we receive per unit area.

1.3 *The Earth*

Among the planets circling the sun is the earth, a solid body which is almost a perfect sphere. Its polar radius is 6357 km, and equatorial radius 6378 km. Its movement around the sun once a year combined with the tilt of its axis produces the changes in the heat reaching any particular latitude. As a result, the equatorial regions have their wet and dry seasons, and the higher latitudes their cold and warm seasons. The daily rotation of the earth about its axis gives day and night, except in the vicinity of the poles. Both movements, therefore, have effects on the weather over the earth.

Of particular significance in meteorology is the rate of rotation about the axis because it affects the winds (see Section 7.5). The rate of rotation is 360° in one sidereal day of 23 hr, 56 min, 4 sec. This gives a rate of rotation of 4.178×10^{-3} degrees per second or 7.29×10^{-5} radians per second.

The other feature of the earth that is of major significance in meteorology is the variation in topography and in the characteristics of the surface layer. The oceans cover 70 per cent of the surface. This is unequally divided between hemispheres, with the Southern Hemisphere being about 80 per cent water and the Northern Hemisphere 60 per cent. The variations in the surface of the land result in variations in weather. Mountain areas produce changes in the air movements and with them come changes in weather patterns. Over the more level areas, the differences in soil, vegetation, and water areas are results of the climatic patterns, but they in turn exert an influence on the weather.

1.4 *The Atmosphere*

The gaseous envelope about the earth is formed by a mixture of gases. With reference to the weather, the most significant one is water vapor. Its changes and the various forms in which water is found are the subject of most of the study by meteorologists.

By considering only dry air, the proportions of the different gases at ground level remain approximately constant both in space and in time. The proportions by volume are given in Table 1.1. These propor-

TABLE 1.1 Fixed Gases in the Atmosphere with Their Proportions by Volume (Per Cent)[a]

Nitrogen	78.08
Oxygen	20.95
Argon	0.93
Carbon dioxide	0.03
Neon	1.82×10^{-3}
Helium	5.24×10^{-4}
Kripton	1.14×10^{-4}
Xenon	8.7×10^{-6}
Methane	1.5×10^{-4}
Nitrous oxide	5×10^{-5}
Ozone	10^{-5} to 10^{-6}
Water vapor	Variable

[a] From Ratcliffe, J. A., 1960, *Physics of the Upper Atmosphere*. New York, Academic Press, pp. 26–27.

tions vary only slightly in the lowest 100 km. Notice that nitrogen and oxygen compose 99 per cent of dry air, and that these gases plus argon add up to 99.96 per cent. The molecular weight of dry air is 28.97.

Other than water vapor, two gases, carbon dioxide and ozone, influence weather processes, and the proportions of both gases vary slightly. Carbon dioxide leaves and enters the atmosphere through the formation and decay, respectively, of organic matter. The oceans by absorbing or releasing the gas help to keep the proportion in the free atmosphere nearly constant. As indicated in Chapter 5, the gas absorbs long-wave radiation from the earth's surface. Variation in the amount would be reflected in changes in the retention of heat by the surface layers, the "green-house effect," with resultant changes in the temperature. Its effect in producing a change in climate is discussed in Section 13.7.

At the earth's surface, the proportion of ozone to the total air mass is 10^{-8} to 1, but this ratio increases to 7×10^{-6} to 1 at an altitude of 20 to 33 km. Above this level, the proportion again decreases. Ozone is formed from oxygen by ultraviolet radiation at the top of the stratosphere (see Section 2.7), resulting in a warm layer. As ozone sinks below the level of maximum concentration, dissociation is not balanced by formation, and so the proportion falls. As pointed out in Section 5.4, the ozone in the stratosphere effectively intercepts the insolation below the wavelength of 0.3 microns,[1] so that little reaches the surface.

Above a cold layer at 80 km the air is very thin. It is thought that here the proportions are different from the ones found near the surface. Observations on the aurora indicate that in these layers oxygen tends to exist in an atomic form.

1.5 *The Heat Balance*

The solar heat that the earth intercepts does not remain. Some of this heat is not even absorbed by the earth or its atmosphere, but is reflected by the clouds, the snow, and other parts of the earth. It is this reflected part that makes our planet visible to a traveler in space. Even the part of the solar energy that is absorbed must be lost again to outer space. Otherwise, the average temperature over the earth would increase from year to year. During historical times, the mean temperature at any one location has varied only slightly and, therefore, we must assume that incoming heat from the sun and outgoing heat from the earth must, in general, balance.

Although, on the average, the energy balance between the planet and outer space is maintained, this is not necessarily true at any one time or at every location. In each hemisphere, spring is the season when income exceeds outgo, and the whole hemisphere warms, while the opposite hemisphere experiences autumn, with a heat debit and tempera-

---

[1] A micron, μ, equal to 10^{-4} cm, is frequently used as a unit of length when the distances are very small.

ture decrease. When the totals for the year are calculated, many areas of the earth do not show a balance between the heat gained from and that lost to outer space. The different altitudes of the sun and the differences in the amount reflected are major causes for differences in absorbed energy.

Incoming heat from the sun is evenly distributed over areas perpendicular to the sun's rays, but it is unevenly distributed when one considers the earth's surface. Directly under the sun each square centimeter intercepts 2 cal min^{-1} (see *AB* of Figure 1.1). On the dark side of the earth, no heat is received directly from the sun. At a place where the sun is just above the horizon, as at sunset or in the polar regions (see *CD* in Figure 1.1), the heat flowing through a unit area perpendicular to the sun's rays is spread over a much greater area at the surface of the earth. Thus the total incoming energy varies with latitude.

The amount of heat lost without being absorbed by the earth also changes with latitude because of different solar elevations and different surfaces. For example, snow reflects the heat while a green field absorbs most of the heat that it receives. The difference, which is the heat absorbed by earth, is given in Figure 1.2. In the diagram, the abscissa is proportional to the sine of the latitude so that areas under the curve are proportional to amounts of heat actually absorbed by the earth. We observe that the average heat absorbed drops from about 570 cal cm^{-2} day^{-1} in the tropical regions to 120 cal cm^{-2} day^{-1} at the poles. The calculations for this diagram were done by Houghton. Other scientists have obtained somewhat different figures, but the conclusions relating to energy flow are much the same as the ones given in the following sections.

The heat lost to outer space varies less with latitude than the heat gained from the sun. This is shown in Figure 1.3, which gives the heat loss as determined by measurements made from satellites. There is variation with the seasons, but the differences are minor. In the zone between

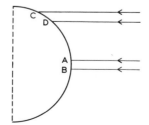

FIGURE 1.1 Effects of the curvature of the earth on the distribution of the sun's heat. Notice that the same amount falls on *AB* and *CD*.

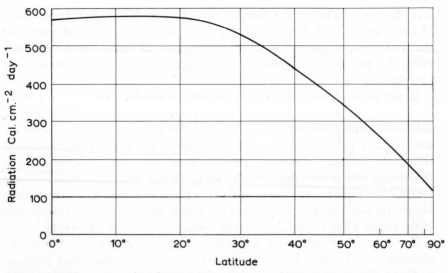

FIGURE 1.2 Mean annual radiation from the sun absorbed by the earth. (From Houghton, Henry G., 1954: On the annual heat balance of the Northern Hemisphere. *J. Meteor.*, 2, 7. Published by permission.)

30°N and 30°S, the net loss is about 510 cal cm⁻² day⁻¹, thus, about 60 cal below the solar heat retained, as shown in Figure 1.2. The loss of heat near the North Pole is about 350 cal cm⁻² day⁻¹, approximately 230 cal cm⁻² day⁻¹ more than that retained from the sun. The loss of heat from the South Polar plateau is even greater. The two flows of heat are approximately equal at 40° lat.

1.6 *Lack of Heat Balance*

The lack of heat balance discussed in the preceding section has a marked significance on the atmosphere. Its magnitude can be better understood if we translate it to other units. The surplus heat of 60 cal cm⁻² day⁻¹ is about the same as would be released if 70 tons of coal were burned every day in every square kilometer (200 per square mile) of the equatorial regions.

To maintain a uniform temperature in the tropics, this surplus heat must be exported to other regions; just as the planet as a whole has a balance between income and outgo, so must there be a balance for every part of the earth's surface. For the same reason, the debit balance of the polar regions must be matched by a transfer of heat from other areas, so that the area does not become colder than it already is. The surplus solar heat that is retained in the zone between 40°N and 40°S must be exported to make up the deficit balance for the areas poleward

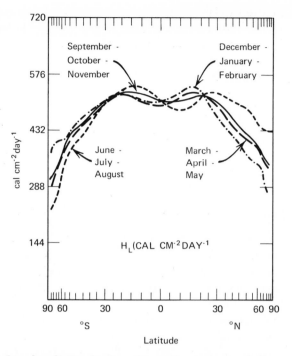

FIGURE 1.3 Infrared radiation leaving the earth as measured from meteorological satellites, averaged along latitudes. (From T. H. Vander Haar, 1967. Published by permission. From Bryson, Reid A., 1968: Other factors being constant . . . *Weatherwise*, 21, 57.)

of 40°. The maximum transport must occur at the latitude of balance, approximately 40°N or 40°S. The heat transport across these parallels has been computed to be 11.2×10^{19} cal day^{-1}, or the equivalent to the solar heat stored in 1.4×10^{13} kg of coal every day.

1.7 *Heat Transport by the Oceans*

One means of transporting heat from the equatorial to the polar regions is provided by the currents of the oceans. In the North Atlantic, for example, the Gulf Stream flows northward from the Caribbean Sea along the east coast of the United States. It leaves the tropics as a warm current, with surface temperatures about 23°C (80°F). As it moves northeastward, part of it becoming the North Atlantic Drift of northwestern Europe, it delivers its surplus heat to the air above and to the coastal areas near it. The warmth of Norway, abnormal for a country so far north, comes mainly from the heat transported there by the ocean currents. The compensating currents are the cold Labrador Current along the Labrador coast and around Newfoundland, and the cool Canaries Current moving southward past the Iberian Peninsula and northwest Africa. As the water moves equatorward, it cools the surroundings as it absorbs heat to replace that which is lost on its journey poleward.

Similar flows, large eddies in the ocean waters, are found in most oceans. By their movements, they assist in carrying the surplus heat from the tropics toward the colder areas of the earth's surface. Extensive as these currents are, they carry only about one third of the surplus energy of the tropics away from these warm areas.

1.8 *Role of the Atmosphere in Heat Transfer*

The other major, and by far the more important, method by which heat is transferred is found in the atmosphere. Part of the effectiveness of the atmosphere in distributing heat can be observed by contrasting temperature extremes on the earth with the ones on the moon where little or no gas is found to carry heat from one part of the surface to another. The hottest regions of the earth, based on the mean annual temperatures, are found in the Sahara with some stations having a mean annual temperature of 29°C (85°F). The highest air temperature recorded under standard conditions, 134°F (57°C), was observed in Death Valley, California, on 10 July 1913. Temperatures slightly above 57°C have been reported, but there has been doubt about the conditions under which they were observed. The coldest spot thus far observed is at station "Vostok" in Antarctica (78°27'S, 106°52'E, elevation 3,488 m). Here a five-year average shows a mean annual temperature of −33°C (−27°F) and a record minimum of −87.4°C (−125.3°F) on 25 August 1958. The official temperatures are taken in a thermometer screen at a height of approximately 4 ft above the ground. The surface temperature may, under extreme conditions, differ by 10 to 15°C, either warmer or colder, from the screen temperature. Even so, the extremes on the earth are much less than on the moon where the surface temperature under the direct rays of the sun rises to more than 100°C, and on the dark side of the moon falls to −150°C.

It is the wind that transports the energy from regions where there is a surplus to other regions where there is a deficit. South winds in the Northern Hemisphere are usually warm because the air has been heated over more southerly latitudes. Passing poleward, they give up some of this heat to the surrounding bodies, warming them. On the other hand, north winds are cold as the air again absorbs heat from the earth's surface, replacing the heat that it lost as it moved into the polar regions.

1.9 *The Rotation of the Earth and Heat Transfer*

Air, like the ocean, transports surplus heat from the tropical regions to the areas where there is a deficit in the radiative heat exchange. There must be a returning current of air to fill the space left empty by the poleward-moving current, just as there is a returning current of water in the oceans. These currents need not be at the surface of the earth or the ocean, but are found at all levels of the atmosphere and ocean, respectively. A simple flow pattern would be present if the air, having been warmed near the equator, rose and at upper levels moved poleward, giving out its surplus heat as in went. Subsiding near the poles, it would follow the earth's surface back to the equator.

As we shall learn in Chapter 7, this simple flow pattern fails because of the earth's rotation. Instead the atmosphere becomes organized in eddies of various sizes. Small eddies, such as whirlwinds and tornadoes, play a negligible part in the latitudinal movement of heat, although they do move heat from the bottom of the atmosphere to upper, colder regions. The transfer by tornadoes is to levels over 10 km above the surface of the earth.

Larger eddies, with diameters 500 km and over, are the major means of the latitudinal transport of heat. Imagine an eddy with a counterclockwise flow of air having its center in the Mississippi Valley and covering most of the United States east of the Rockies. In the east, southerly winds will carry air warmed over the Gulf of Mexico northward toward the Great Lakes, warming the eastern states in its flow. To the west of the Mississippi, cold air will flow from the Canadian prairies, chilling inhabitants as it acquires heat that it will again lose in another movement poleward.

The weather map provides a picture of these eddies in the atmosphere and their changes from day to day. An understanding of these various eddies, their causes and their effects, is the aim of the meteorologist, and an introduction to their study is found in the subsequent chapters of this book. It is by means of these larger eddies that surplus heat from the tropics is transferred to polar areas and, as a result, the range of temperature over the earth is kept as small as it is.

1.10 *Water and the Heat Balance*

Another important, but less obvious, factor in balancing the heat budget of the earth is the water found in the atmosphere. Water is always present in an invisible form as water vapor and sometimes as small droplets or ice crystals forming the clouds and fog that are found over the earth, and frequently falling as rain or snow or hail. The moisture in the atmosphere is discussed in detail in Chapter 4.

Water enters the atmosphere from the surface of the earth. A small amount is formed in the combustion of hydrocarbons, but most of it evaporates from water surfaces or from moist soil, snow and ice surfaces, or the leaves of plants. The change from liquid to vapor or solid to vapor requires heat, approximately 600 cal gm^{-1} and 678 cal gm^{-1}, respectively. This stored heat, or *latent heat*, is released when the reverse process occurs and the vapor once again turns to the liquid or solid form. Thus, water vapor provides a storage bank for retaining and later releasing heat. To distinguish it from latent heat, the heat energy that is measured by a thermometer is sometimes called *sensible* heat, or the heat that one can detect with the senses.

The greatest amount of evaporation occurs from the oceans and the tropical forests near the equator. In this evaporation, the water is using and storing some of the surplus solar heat of the Equatorial Zone. Much of the moisture condenses into clouds over these same areas, releasing the stored or latent heat, several kilometers above the surface. Yet some of this vapor is carried poleward by the air currents near the earth's surface, later to condense into clouds with the release of its latent heat. By this means, movement of moisture contributes to the total heat transfer.

The magnitude of the effect can be realized by considering the heat released before 10 cm of rain can fall on 1 km^2. The volume of rain is 10^{11} cm^3 and, hence, the weight is 10^{11} gm. The heat released as the vapor turns to liquid is 6×10^{13} cal, equal to the heat released with the burning of approximately 8 thousand tons of coal. The release of this heat is not apparent to the people on the ground because it warms the air 2 km above them, where the water vapor turns to liquid or solid.

The magnitude of the transport of energy across specific latitude circles of the Northern Hemisphere, for the year 1950, are given in Table 1.2, both for sensible and for latent heat. Notice that the proportion of the total amount transferred as latent heat drops from $\frac{1}{2}$ at 31°N to $\frac{1}{5}$ at 70°N. In higher latitudes, the air is too cold to carry much water vapor.

As stated in the preceding section, the eddies that transport the sensible heat and the latent heat poleward from the tropics are large, but they are not large enough to make the transfer in one step. The current in the example of Section 1.9 will carry moisture evaporated from the

Gulf of Mexico northeastward to the lower Great Lakes where it will fall as rain. Later, this moisture will again evaporate and be caught up in another eddy to be carried farther north. Thus the transfer is similar to that of a bucket brigade, with some of the heat being intercepted by the intermediate latitudes.

TABLE 1.2 Transport of Energy in Units of 10^{19} Cal Day^{-1} across Specified Latitude Circles (after Starr and White)[a]

Latitude (°N)	Sensible Heat	Latent Heat	Total
70	2.9	0.7	3.6
55	5.1	2.0	7.1
42.5	4.9	2.4	7.3
31	2.4	2.7	5.1

[a] Starr, V. P., and R. M. White, 1954. Balance requirements of the general circulation. Cambridge, Mass. Air Force Cambridge Research Center, Geophysical Research Papers No. 35, 62 pp.

This chapter has shown that the wind and the weather that every one experiences are a small part of a whole which encompasses the globe. The total picture is complex, and the interrelationships are many and can be far-reaching. Because of the flows in the atmosphere, the variations in meteorological conditions on the surface of the earth are small compared to variations in outer space. Under these conditions the human race has been able to adapt itself so that man can survive on most parts of the land areas of the globe. The succeeding chapters provide an introduction to the complexities of the flow and the causes for them.

PROBLEMS AND EXERCISES

1. In Section 1.6, the surplus in the radiative transfer of heat at the equator is given as 60 cal cm^{-2} day^{-1}. One pound of coal releases approximately 3.5×10^6 cal. Confirm the figure of 70 tons of coal, given in the section.

2. Verify the figure of 8000 tons of coal mentioned in Section 1.10.

3. One gram of TNT releases, on explosion, 1280 cal. Compare the energy release of a megaton bomb (equivalent to 1 million tons of TNT) and the energy released from condensation in a storm that gave, on the average, 5 cm of rain over 2×10^4 km^2.

CHAPTER 2

TEMPERATURE

2.1 *Temperature*

An important variable in discussions about the weather is the temperature. The physicist defines temperature in terms of the movement of the molecules—the more rapid their motion, the higher the temperature. The layman prefers terms such as hot, mild, and bitterly cold in describing the weather, even though he has no precise definition of temperature. These terms do not have the same meaning for all people, nor even for the same man in different seasons of the year. A mild day in winter would, in many regions, be called a very cold day in summer. Therefore, some standard must be set so that temperatures may be evaluated and compared.

2.2 *Thermometers*

The instrument used to measure temperature is called a *thermometer*. There are several types of thermometers, the design varying with the purpose and with the method of use.

All thermometers measure temperature and its changes by using the effect that heat has on a body. The expansion that occurs with the addition of heat is the one most commonly used. The expansion of mercury contained in a sealed glass tube gives the most common scien-

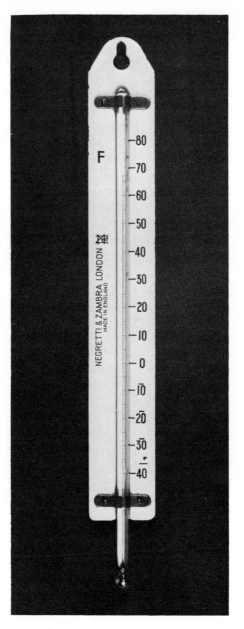

FIGURE 2.1 Mercury thermometer. (Meterological Service of Canada.)

tific thermometer, the mercury-in-glass or *mercury thermometer* (Figure 2.1). Sometimes alcohol is substituted for mercury because the latter solidifies at —39°C. The relative expansion of two metals with changing temperatures provides the key to the bimetallic thermometer. Use is made of a bimetallic thermometer in one design of a *thermograph* (Figure 2.2) where the variations in temperature are recorded on a rotating drum.

Changes in temperature cause changes in the resistance of a conductor of an electric current, permitting the development of the *resistance thermometer*. Also, a *thermocouple* measures temperature because of the change in the electromotive force at a junction of two metals. The amount and character of energy emitted by a body through radiation (to be discussed in Chapter 5) can be used to measure the temperature of its surface. The *radiation thermometer* is less simple than the other thermometers, but it has the advantage of being able, under certain circumstances, to measure temperature at a distance. Thus, one may

FIGURE 2.2 Thermograph, barograph, hydrograph. (Meteorological Service of Canada.)

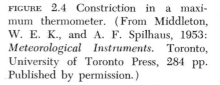

FIGURE 2.3 Index in a minimum thermometer. (From Middleton, W. E. K.,
and A. F. Spilhaus, 1953: *Meteorological Instruments.* Toronto, University of
Toronto Press, 284 pp. Published by
permission.)

FIGURE 2.4 Constriction in a maxi-
mum thermometer. (From Middleton,
W. E. K., and A. F. Spilhaus, 1953:
Meteorological Instruments. Toronto,
University of Toronto Press, 284 pp.
Published by permission.)

be put on a satellite circling the earth and used to measure the temperature of the cloud tops or the surface of the earth. The minimum and
maximum thermometers are two types of thermometers designed to meet
special needs and much used by climatologists. An index in an alcohol
thermometer (Figure 2.3) moves downward by the surface tension of
the alcohol and remains in its lowest position with rising temperatures
as the alcohol flows past the index. A constriction in a maximum thermometer (Figure 2.4) holds the mercury at its highest point, thus
recording the maximum temperature until it is forced back past the
constriction when the observer shakes the thermometer.

2.3 *Temperature Scales*

In order that temperatures may be compared, it is necessary to have
standards of measurement. Three scales of temperature, the Celsius or
Centigrade, the Fahrenheit, and the Kelvin or Absolute, are currently
in general use. These scales are indicated by C, F, and K or A,
respectively.

The basic points that determine the values are the temperature of
a mixture of ice and water (the melting point of ice) and the temperature of steam above boiling water at sea level (the boiling point of
water). These points are, respectively, on the Celsius scale, 0°C and
100°C, on the Fahrenheit scale, 32°F and 212°F, and on the Kelvin
scale, 273°K and 373°K. To change from the Kelvin scale to the Celsius
scale is simple—it is done merely by subtracting 273. The degree is the
same size on the two scales.

To change from Fahrenheit to Celsius is more difficult. The difference
in temperature between the freezing point and the boiling point of water
is 180 Fahrenheit degrees, or 100 Celsius degrees. Thus the sizes of
the degrees are in the ratio of 5:9. Knowing this, one can begin by

using the relationship $0°C = 32°F$ to make a table of corresponding values.

°C	−40	−20	−10	−5	0	5	10	15	20	30
°F	−40	−4	14	23	32	41	50	59	68	86

From this table, one can readily transfer from one scale to an approximate value on the other scale. The exact relationship may be obtained from

$$(T − 32)°F = \tfrac{9}{5}(T − 0)°C$$

which becomes

$$T°F = \tfrac{9}{5}T°C + 32 \quad \text{or} \quad T°C = \tfrac{5}{9}(T°F − 32)$$

The Celsius or Centigrade scale is the one in general use in most countries of the globe and in all upper-air observations. Also the exchange of weather data is made by using the Celsius scale. The general public of the United States and Canada still uses the Fahrenheit scale, and so surface temperatures are still reported in Fahrenheit in these countries. The Celsius scale, or the Kelvin scale readily obtained from it, has been the basis for scientific definitions and computations and, hence, its use is becoming gradually more widespread. Consequently, it will be used generally in this book.

2.4 *The Measurement of Air Temperature*

One of the difficulties facing a meteorologist or climatologist is to know what is meant by the "air temperature." Even during a stroll, one can become conscious that there are variations in the temperature of the air surrounding him. It is warm on the sunny side of a building, and cool on the shady side. Within a forest there are pockets of mild air in contrast with the general coolness. Careful measurements confirm that the temperature changes from place to place. These changes are associated with the underlying surface, the topography, the height above the surface, the time of day, and other factors.

This problem is indicated in Figure 2.5, which shows temperatures recorded at nearby sites by different organizations, at O'Neill, Nebraska, on 22 August 1953. (Notice that the height scale is not linear but logarithmic.) At 0435 h, the range of temperature was from 13.9°C at 0.1 m to 16.5°C at 16 m. At 1235 h, the range was from 30.5°C at 0.1 m to 26.9°C at 16 m. Even at the same heights, the daytime observations showed rapid swings of temperature. Six-second observations at heights from 0.5 to 4.0 m during a 10-minute period centered at 1235 h showed

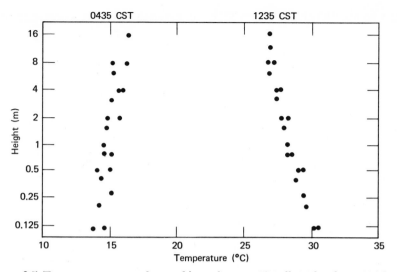

FIGURE 2.5 Temperatures near the earth's surface at O'Neill, Nebraska, 0435 h and 1235 h CST 22 August 1953.

swings of more than 3°C at every level. At the same levels, at 0435 h, most of the observations were within 0.1°C of the mean.

Figure 2.6 illustrates the effect of the underlying surface. The diagram shows the temperature distribution at Hamilton, at the western end of Lake Ontario, at 2200 h EST 24 October 1966. As pointed out in the diagram, the water temperature in the protected Hamilton Harbour was 12°C, although in the open lake it was 6°C. At the time of observation, there was a warm spot, temperature 11°C, on the bank at the eastern end of the harbour, a result of the warm water. Within the city, there were three separate warm spots, found in centers of high building density. In contrast with these warm areas, cool areas were found along the southern boundary of the city and in the valley of Dundas Creek on the southeast corner of the city. The latter cool spot arose because of the drainage of cool air down the sides of the valley (see Section 7.13). The temperature gradient along the southern boundary of the city reached a value of 3.8°C km^{-1} (11°F mi^{-1}). Other observations in other areas have shown that, particularly on a calm night, temperature gradients develop, as in Hamilton, because of urban-rural contrasts, because of water-land contrasts, and as a result of topography.

Temperature variations with distance may be determined very rapidly by means of radiation thermometers mounted on aircraft. Figure 2.7 shows the results obtained during a flight over Lake Ontario on 3 May

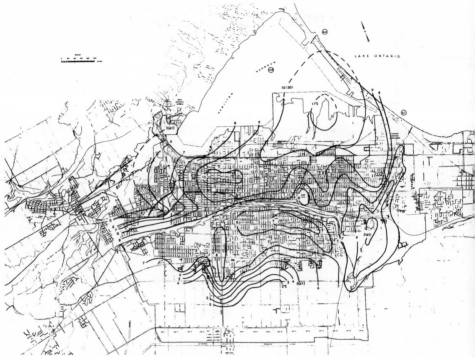

FIGURE 2.6 Distribution of temperature in Hamilton, Ontario at 2200 h EST 24 October 1966. Isotherms are numbered in degrees Celsius with some Fahrenheit temperatures given in parentheses. Water temperatures are encircled. (From Oke, T. R., and F. G. Hannell, 1968: The form of the urban heat island in Hamilton, Canada. World Meteor. Organization, paper presented to the symposium on Urban Climates and Building Climatology, Brussels, Belgium, 16 October 1968. Published by permission.)

1965. The record does not give the air temperature, but instead the temperature of the surface, water or land, over which the aircraft was passing. Therefore, the results are not directly comparable with the ones illustrated in Figures 2.5 and 2.6 but, because the air tends to adjust its temperature to that of the underlying surface, the results are of interest. The open lake showed temperatures in the range 4 to 8°C. Near Grimsby, on the south shore, the water temperature was warmer, reaching 12°C at the coast. Near Clarkson on the north shore, the change to warmer waters near the coast was more abrupt. The changes over the lake, although significant, were minor compared to the ones over the land. Around Grimsby, the temperature fluctuated between 25°C and 31°C, with one spot having a temperature of 38°C. This latter spot was probably part of some building used for heating. Clarkson,

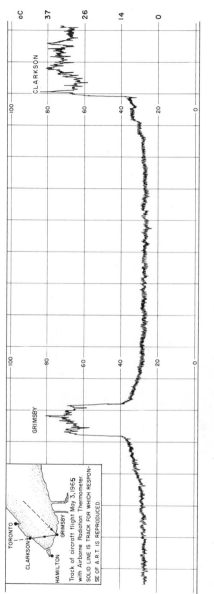

FIGURE 2.7 Temperature observations obtained by a radiation thermometer carried on a flight over western Lake Ontario 3 May 1965. (Courtesy, Meteorological Service of Canada.)

19

more thickly populated than Grimsby, had even higher and more variable ground temperatures than Grimsby.

The results of studies like the ones illustrated in Figures 2.5, 2.6, and 2.7 have shown that the determination of a unique "air temperature" is not simple. For purposes of record and comparison, official air temperatures are taken within a white louvred screen (Figure 2.8) at a height of approximately $1\frac{1}{2}$ m (4 ft). Usually, the thermometer is ventilated by a fan which draws air past it. Whenever possible, the screen is located over a plot of grass that is kept cut. From our description above, it is to be expected that other temperatures close by can differ considerably from the "official" temperature. This difference is particularly significant with frosts. Temperatures readily go below freez-

FIGURE 2.8 Thermometer screen showing thermograph and maximum and minimum thermometers. (Courtesy, Science Associates.)

ing at the ground when the minimum temperature within the screen is 38°F (3°C), and the difference between the "official" temperature and the one in a neighboring valley can be as much as 15 or 20°F.

2.5 *Temperature Gradients*

The rate of change of temperature with distance, illustrated in Section 2.4, is called the *temperature gradient*. The term *gradient* has been adapted from surveying where it refers to the rate of increase of height with distance. In meteorology, it is loosely used to indicate the magnitude of the change of a variable with distance, without regard to sign. Thus one speaks of the pressure gradient or the moisture gradient or the temperature gradient.

The vertical temperature gradient has been called the *lapse rate*, but with this term the direction in which the temperature increases is specified. A positive lapse rate means a decrease of temperature with height. A negative lapse rate is called an *inversion*. As shown in Figure 2.5, there was an inversion near the ground at O'Neill at 0435 h, 22 August 1952.

An idea of the magnitude of horizontal temperature gradients may be obtained from Figure 2.9. On it are given the temperatures and winds for 1200 h Greenwich Mean Time (GMT) or 0700 h EST 8 January 1969 over much of North America. The system of plotting is the one usually used on weather maps. The temperature is given in degrees Fahrenheit to the upper left of the station circle. The wind direction is given by the "shaft" of an "arrow," the wind blowing along the shaft toward the station. The speed is given in barbs and half-barbs, a barb representing approximately 10 miles per hour (written 10 mi hr^{-1}) or 4 m sec^{-1}. When there is no wind, a second circle is drawn around the station circle.

Lines of equal temperature, called *isotherms*, are drawn on Figure 2.9 at intervals of 9°F or 5°C. These permit a rapid evaluation of the temperature gradient. Where the isotherms are far apart, as in southwestern United States and north central Canada, the temperature gradient is weak. In contrast, there were bands through the Dakotas and Montana and through Kansas and Colorado where the gradient approached 1°F in 8 mi, or 1°C in 20 km.

The vertical temperature gradient is usually much greater than the horizontal temperature gradient. This is shown on Figure 2.5, where the lapse rates were —2°C in 16 m at 0435 h (an inversion), and 3°C in 16 m at 1235 h, or in English units —1°F in 15 ft and 1°F in 10 ft, respectively. These rates are not abnormally high. Differences of 6°C

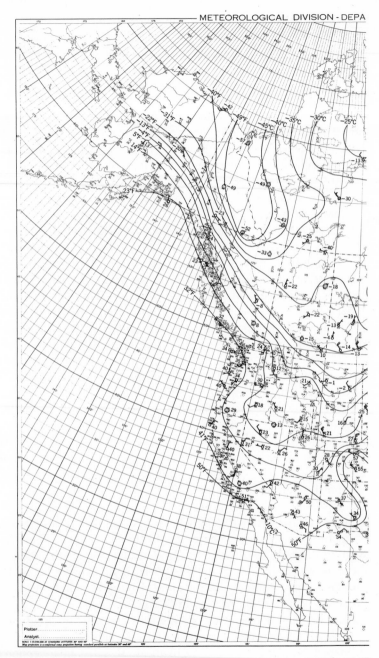

FIGURE 2.9 Temperatures and winds over a portion of North America at 1200 h GMT the station circle. The wind blows along the shaft toward the second circle indicates a calm.

22

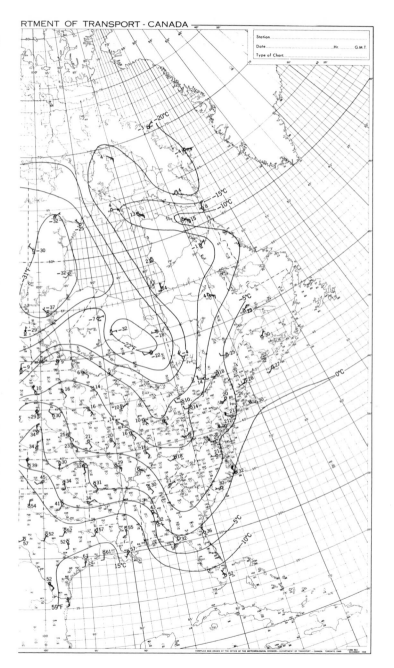

(0700 h EST) 8 January 1969. The temperature is plotted in degrees Fahrenheit near station circle, and the speed is given by the relationship: 1 barb $\simeq$ 10 mi hr^{-1}. A

23

have been found between 2.5 cm and 30 cm. The strongest inversions are found above snow surfaces with light winds. Under abnormal circumstances, a rise of 40°F in 40 ft can be found. Extreme values of positive lapse rates are found above desert sands or above asphalt on a sunny afternoon with light winds. Extreme values are seldom found in the free air.

2.6 *Temperature Changes*

At many stations throughout the world, temperature observations are made under conditions established by national weather organizations. At some stations, called climatological stations, the observations are of the daily maximum and minimum temperatures only. Other stations take records of the weather elements, including temperature, at the four *synoptic hours*, that is, 0000, 0600, 1200, and 1800 h, GMT. Reports are collected by telegraph and are used to prepare the weather maps (see Chapter 10). Other stations, usually associated with aircraft travel, keep hourly records of weather. Examples of hourly weather reports are found in Appendix 4. From all of these records, much can be learned about the temperature changes in both space and time.

The daily variation of temperature is familiar to all. Figure 2.10 gives the thermograph and hygrograph (see Section 4.3) record at Ellerslie, near Edmonton, Alberta, for the week 6 to 13 May 1968. The temperatures are in degrees Fahrenheit; the lower graph shows the relative humidity (see Section 4.2). On every day except May 9, there was a relative maximum temperature during the afternoon, usually between 1400 and 1700 h. May 9 was different. It rained during the early morning. Then it cleared by 0800 h, and a brisk north wind, 30 mi hr^{-1} or more with gusts up to 50 mi hr^{-1}, caused the temperature to drop from 43°F at 1000 h to 33° by mid-afternoon. The slight range of temperature, 6°F, on May 6 was because of an overcast sky. In contrast, May 11 with sunny skies and light winds saw the temperature rise from 30° to 72°. Minimum temperatures usually occurred between 0400 and 0500 h, or close to the time of sunrise at 0440 h.

The most significant change in temperature in the tropical and equatorial regions is the diurnal change, but in the higher latitudes the annual change has greater significance. Normally, the hottest time of the year comes about 4 weeks after the summer solstice, and the coldest time of year about 4 weeks after the winter solstice. Deviations of more than 3 weeks from these times are abnormal, although they do exist. This topic is discussed further in Section 5.6. Large differences in the magnitude of the annual range are easily found. For instance, Fairbanks,

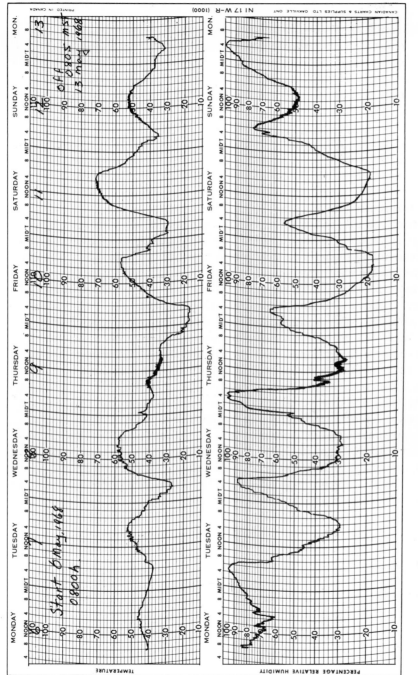

FIGURE 2.10 Thermograph record for Ellerslie, Alberta (53°N 114°W) for the week 6 to 13 May 1968.

Alaska with an average January temperature of —10°F, is 36° colder than Juneau, in the Alaskan panhandle. In contrast, Fairbanks in July with a mean of 61°F is 6°F warmer than Juneau. The change from the coldest to the warmest month, that is, *the annual range of temperature*, at Juneau of 29° is less than one half the annual range at Fairbanks of 71°.

The diurnal and annual variations in temperature are closely associated with the movement of the earth on its axis and around the sun. The solar influence is direct, but other factors modify its effects on the ranges of values. Other changes, outbreaks of cold polar air and waves of mild air, for example, are not rhythmical. The weather map, to be discussed in Chapter 10, but also mentioned in other places in the text, has proved useful in explaining these random changes. Because these variations are irregular in both time and degree, they and their accompanying changes in precipitation, wind, etc., provide most of the interest in the weather.

Figure 2.9 provides an example of these random variations in temperature. If the motion and altitude of the sun were the only causes for temperature changes, these isotherms should follow the parallels of latitude closely, with some adjustment because of differences in local time. High temperatures would be near the equator and low near the poles. Figure 2.9 shows that this was generally true at the time of the map. The warmest areas, with temperatures above 60°F, were near the Gulf of Mexico, and the values dropped as one went northward. But other factors entered so that isotherms did not run along east-west lines. Thus, at approximately 44°N latitude, Portland, Maine was 28°F and Toronto, Ontario, 10°F. Also, at 36°N, Knoxville, Tennessee was 21° and Amarillo, Texas, 54°. Over much of the continent, winds were light. Strong north winds were bringing cold air into the western plains. In the warm air on the Texas coast, the winds were from the south or southwest.

From individual thermometer readings, mean values are obtained. For stations with continuous weather observing, the mean daily temperature could be taken as the average of the 24 hourly readings. Many climatological stations record the maximum and minimum temperatures only. From them, the *mean daily temperature* is obtained by taking the average of the maximum and minimum. Often, the same method is used even when more frequent observations would permit a more accurate mean. Mean monthly temperatures are determined from the mean maximum and mean minimum. The yearly mean is the average of the monthly values, ignoring the varying lengths of the months. When a station has been taking records for a number of years, normal values can be obtained by averaging the yearly values. Figure 2.11a, b gives the normal January temperatures for the United States and Canada, and Figure 2.12 shows the normal July temperatures. These months are chosen because for most stations they represent the coldest and the warmest months, respectively.

The isotherms on Figures 2.11 and 2.12 show that the averaging process eliminated many of the random variations in temperature noted above. Other effects on temperature, in particular the effect of water areas and ocean currents, have modifying influences on the solar control and cause the isotherms to deviate from the parallels of latitude.

The influence of altitude on temperature in mountainous areas is so great that normal values cannot be used in small scale maps and produce meaningful results. For these maps, the normal values are "adjusted to sea level" to provide a common base for comparison and to bring out other pertinent facts. This is accomplished by adding 1°F to the temperature for every 330 ft altitude (see Problem 2). These adjustments have been made for Figures 2.11 and 2.12.

This effect of altitude on temperature is shown in Figure 2.13. Here are given the mean maximum July temperatures for Idaho for the period 1931 to 1952. The scale is large enough to permit the drawing of actual isotherms based on the observed values. The map shows cold areas for the higher regions with temperatures below 76° and warm areas in the valleys, with a maximum of 96° in the southwestern valleys.

2.7 Upper-air Temperature—Atmospheric Layers

In the preceding section, we observed that higher stations are usually colder, the drop in temperature being a function of height. Scattered observations by balloonists during the 19th century showed that this drop in temperature also occurred in the free atmosphere. Scientists, lacking other information, extrapolated this trend, and assumed that the temperature aloft continued to drop until, at the threshold of outer space, the temperature was close to absolute zero.

This assumption was proved to be false near the turn of the century when recording instruments, borne aloft to great heights by unmanned balloons, were later recovered. They showed that the temperature drop with height ceased abruptly at approximately 10 km. Above this level the temperature remained almost constant with height, at times even rising slowly. The lower atmosphere was called the *troposphere*, the layer above the break, the *stratosphere*, and the dividing surface, the *tropopause*.

The development of the *radiosonde* in the 1920's has made possible a more thorough knowledge of these two layers. A radiosonde is a light-

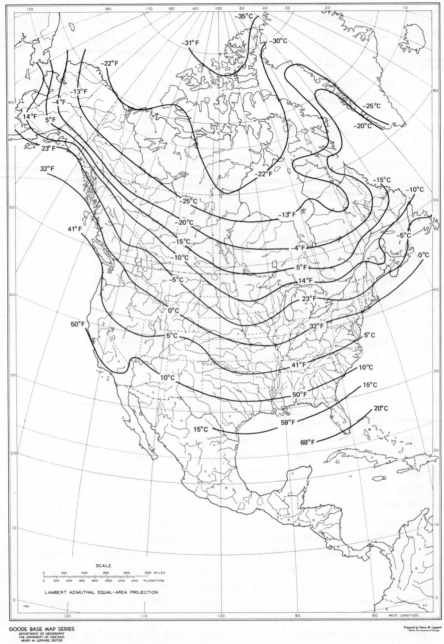

FIGURE 2.11a Mean sea-level temperatures for United States and Canada, January. (In this and in Figure 2.12, the temperatures are adjusted to sea level by adding 3°F for every 1000 ft elevation.)

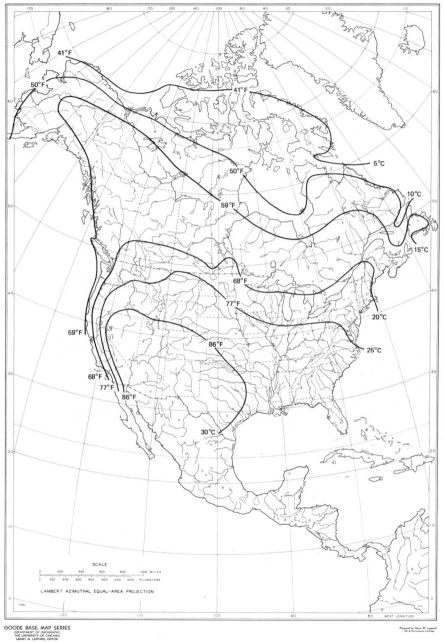

FIGURE 2.11b North America with 3000 ft and 6000 ft contours for United States, Canada, and Mexico.

29

SCALE

LAMBERT AZIMUTHAL EQUAL-AREA PROJECTION

GOODE BASE MAP SERIES
DEPARTMENT OF GEOGRAPHY
THE UNIVERSITY OF CHICAGO
HENRY M. LEPPARD, EDITOR

FIGURE 2.12 Mean sea-level temperatures, United States and Canada, July.

30

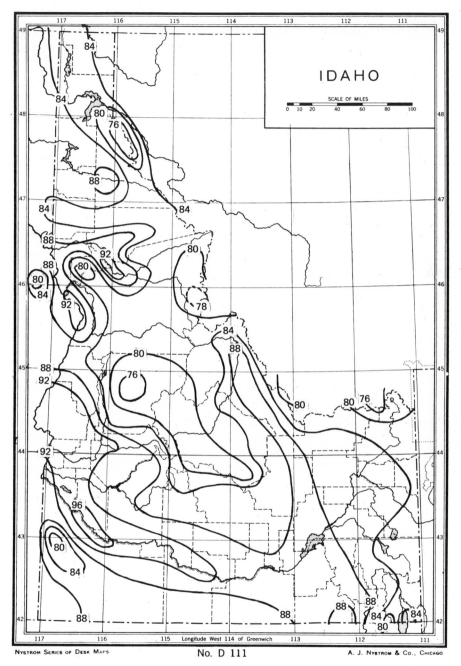

FIGURE 2.13 Mean maximum temperatures (°F), 1930 to 1952 in Idaho, July. (From U.S. Department of Commerce, Weather Bureau, 1959: *Climates of the States, Idaho.* Washington, D.C., 16 pp.)

weight box containing a resistance thermometer, an aneroid barometer (Section 3.2), an electrical hygrometer (Section 4.3), and a radio transmitter. The radio transmits signals that can be interpreted to give the pressure, temperature, and humidity in the atmosphere. Radiosondes are sent aloft by many stations over the earth twice daily, 0000 and 1200 h GMT, to collect information about the state of the atmosphere. They rise usually to about 20 km and frequently to 30 km.

Examples of the temperature distribution in these layers are given in Figure 2.14, which gives the change of temperature with height at 0000 h GMT 17 December 1961, at Swan Island, in the Caribbean; Charleston, South Carolina; Sault Ste. Marie, Michigan; and Fort Smith, Northwest Territories. Tropopauses were found at that time at 7.2 km at Fort Smith, at 12.3 km at Sault Ste. Marie, at 14.0 km at Charleston, and at 16.1 km at Swan Island. At Fort Smith, the temperature was nearly isothermal in the lowest 3 km. The other three stations showed the more normal decrease in temperature with height in the troposphere, although Charleston had a sharp inversion near the surface and the other two showed thin inversion layers.

Figures 2.15 and 2.16 present two cross sections, one for 20 January 1958, the other for 15 August 1957, of the atmosphere from 10°N to 80°N along the 80th meridian, and give more information on the variation of temperature with height. Of course, the vertical scale is considerably expanded in comparison with the horizontal scale. The heavy dark lines indicate the tropopauses, and the lighter lines give the isotherms. Figures 2.15 and 2.16 present upper-air temperatures for individual days, one in winter and one in summer. Some of the variations shown were valid for these specific days only, but most of the characteristics of the temperature distributions are typical of the ones found on most or all days of the seasons they represent.

In all seasons, the tropopause is highest over the tropics. On individual days the surface has breaks, and on occasion there appears to be a double tropopause over an area. Even over those areas where the tropopause is continuous, there are undulations that are smoothed out by taking averages. In general, the tropopause is lowest in the arctic regions. Here, in winter, the surface is usually found between 5 and 8 km; in the summer it is found between 8 and 11 km. At the equator the tropopause is usually at 16 to 18 km, in which region the air is very cold. To appreciate the magnitude of the temperature gradient, notice that above the equator the temperature drops 100°C in an ascent of 17 km (10 mi). The tropopause surface is significant in a study of the weather, since the stratosphere is usually very dry, and the weather phenomena of clouds, fog, precipitation, and dust are restricted

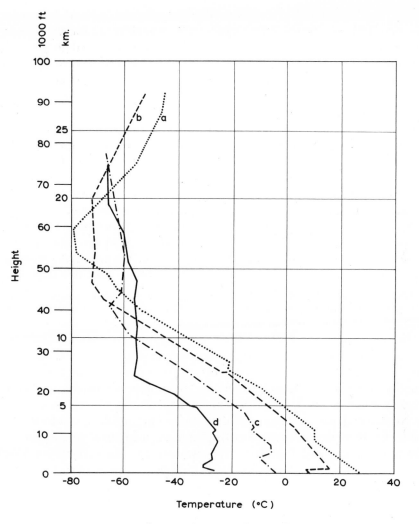

FIGURE 2.14 Temperature soundings in the troposphere and lower stratosphere, 0000 h GMT 17 December 1961: (*a*) Swan Island, West Indies; (*b*) Charleston, South Carolina; (*c*) Sault Ste Marie, Michigan; (*d*) Fort Smith, Northwest Territories.

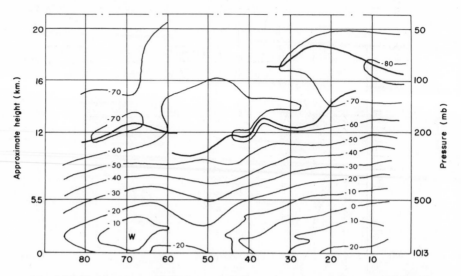

FIGURE 2.15 Temperatures (°C) in a vertical cross section of the atmosphere along the 80°W meridian, 20 January 1958. (From U.S. Department of Commerce, Weather Bureau, 1963: *Daily Aerological Cross Sections, for the IGY period*. Washington, D.C.)

to the troposphere. Variations in the stratosphere have effects on the earth's weather, but in an indirect manner.

Notice that the 20 January observations (Figure 2.15) showed inversions at several places in the troposphere. The ones found north of 50°N are typical of the air over snow surfaces. An inversion at 30°N was associated with the weather situation occurring at that time. Inversions like this one are frequently found, but they move with the weather systems. This type of inversion will be studied in Sections 6.9 and 10.4.

The lower part of the winter stratosphere is relatively uniform in temperature, although even at this time of year the coldest parts are found above the equator. Horizontal gradients are more extreme in summer, with temperatures increasing from the equator poleward to about 60°N. North of this parallel, the temperatures are usually fairly uniform.

Recent probings by means of rockets have given direct information about the temperature above the levels reached by radiosondes. They have shown a temperature distribution as illustrated in Figure 2.17. The isothermal stratosphere gives way at about 20 km to a layer of rising temperatures. At the top of this layer, at about 50 km, the tempera-

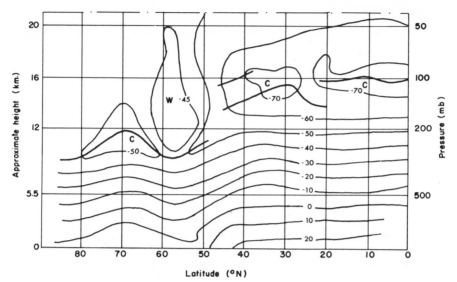

FIGURE 2.16 Temperatures (°C) in a vertical cross section of the atmosphere along the 80°W meridian, 15 August 1957. (From U.S. Department of Commerce, Weather Bureau, 1963: *Daily Aerological Cross Sections for the IGY period.* Washington, D.C.)

ture is again close to freezing. Above, between 50 and 80 km, the temperature drops once again to near 200°K. At 80 km, the lapse rate reverses itself again, and at very high altitudes the temperature is very high. The warm surface at 50 km, which is the top of the inversion, is called the *stratopause.* The layer above is called the *mesosphere,* and its lid, at the point where again the lapse rate reverses itself, the *mesopause.* The warm layer above 80 km is called the *thermosphere.*

These layers and their causes are of interest to students of radiation and its effects on weather. As yet, little is known about possible linkages between the variations at these levels and the weather at the earth's surface. Clouds and weather are phenomena of the troposphere and, therefore, the major interest in this text will be on variations of meteorological elements in the troposphere and lower stratosphere.

For scientific and engineering use, a *U.S. Standard Atmosphere* has been proposed by the Committee on Extension to the Standard Atmosphere. It has been modified slightly since it was originally defined. Its specifications are given in Table 2.1. The lapse rates are assumed to be constant in the intervals between the levels given in the table.

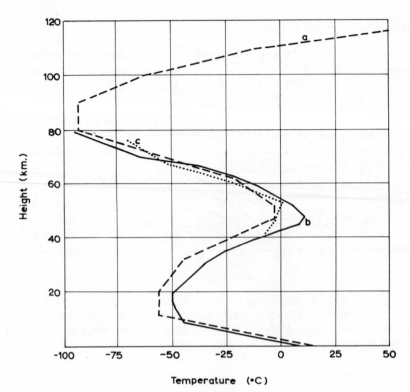

FIGURE 2.17 Temperatures in the high atmosphere: (*a*) U.S. Standard Atmosphere; (*b*) Churchill, Manitoba, summer; (*c*) White Sands, Arizona, September.

TABLE 2.1 Skeleton of the U.S. Standard Atmosphere, 1962

Ht (km)	T (°C)	p (mb)	Density (gm cm⁻³)	Lapse Rate (°C Km⁻¹)
0	15.0	1013.25	1.225×10^{-3}	6.5
11	−56.5	226.	3.64×10^{-4}	0
20	−56.5	54.7	8.80×10^{-5}	−1.0
32	−44.5	8.68	1.32×10^{-5}	−2.8
47	− 2.5	1.11	1.43×10^{-6}	0
52	− 2.5	5.90×10^{-1}	7.59×10^{-7}	1.8
62	−20.5	1.82×10^{-1}	2.51×10^{-7}	4.0
80	−92.5	1.04×10^{-2}	2.00×10^{-8}	0
90	−92.5	1.64×10^{-3}	3.17×10^{-9}	−3.0
100	−62.5	3.01×10^{-4}	4.97×10^{-10}	−5.0
110	−12.5	7.35×10^{-5}	9.83×10^{-11}	−10
120	87.5	2.52×10^{-5}	2.44×10^{-11}	−20
150	687.5	5.06×10^{-6}	1.84×10^{-12}	
190	1078	1.68×10^{-6}	4.35×10^{-13}	
400	1888	4.03×10^{-8}	6.49×10^{-15}	

2.8 *Heat and Heat Units*

In the field of meteorology, as in many other sciences, one deals with the concept of heat. Heat is commonly measured by means of a unit of heat known as the *gram-calorie* or the *calorie*. This is defined as the amount of heat required to raise one gram of water from 14.5°C to 15.5°C. The specific range of temperature is stated because the amount varies slightly with temperature, but for most problems in meteorology this variation may be ignored.

The effect of a given quantity of heat is not the same on all bodies. For instance, if 1 kg of copper at 40°C is dropped into 1 kg of water at 20°C, heat from the copper will flow into the water until an equilibrium temperature is reached. But this is 21.7°, not 30°. The result is related to the different *heat capacities* of the two substances. The heat capacity of an object is the amount of heat required to raise its temperature 1°C. When the object has a mass of 1 gm, the heat capacity is called the *specific heat*, usually abbreviated c. The value of c varies with different substances. In the example, the heat capacity of 10^3 gm of water is 1 cal gm^{-1} $°C^{-1}$ $\times 10^3$ gm or 10^3 cal $°C^{-1}$, and the amount of heat required to raise the water 1.7°C is 1700 cal. The specific heat of copper is 0.093 cal gm^{-1} $°C^{-1}$, so the heat capacity of 10^3 gm of copper is 0.093 cal gm^{-1} $°C^{-1}$ $\times 10^3$ gm = 93 cal $°C^{-1}$. The heat released by the copper is 18.3° $\times$ 93 cal $°C^{-1}$ = 1700 cal. Thus 1700 cal have flowed from the copper to the water, the copper dropping 18.3°C while the same mass of water warmed only 1.7°C.

Specific heats of some substances are given in Table 2.2. Two values are given for air and water vapor because the heat required to raise the temperature of a gas varies with its external conditions. This is treated further in Section 6.1. Commonly used specific heats are those for a gas at constant pressure, and at constant volume. These specific heats are abbreviated, respectively, c_p and c_v. The addition of moisture to dry air changes the specific heats only slightly.

TABLE 2.2 Specific Heats for Common Substances (cal gm^{-1} deg^{-1})

Water	1.00	Copper	0.093
Ice	0.5	Iron	0.107
Water vapor c_p	0.44	Quartz sand, dry	0.19
c_v	0.33	8.3 per cent moisture	0.24
Dry air c_p	0.240	Sandy clay 15 per cent moisture	0.33
c_v	0.171	Wet mud	0.60
		Brick	0.20
		Granite	0.19

In two situations, a flow of heat is not accompanied by a change in temperature. The first occurs when heat is applied to ice at 0°C. This is used to melt the ice, and no rise in temperature occurs until it is completely melted. The heat absorbed in melting amounts to 80 cal gm⁻¹. This heat, called the *latent heat of fusion*, is released when the water freezes. Because of this latent heat, bodies of water near the freezing point, and with them the surrounding land areas, cool slowly. Similarly, when water evaporates, it absorbs heat with no increase in temperature. This *latent heat of vaporization* is released when the vapor turns once again to a liquid. The value of the latent heat of vaporization depends upon the temperature. In the range of normal atmospheric temperatures,

$$L = (597 - 0.56T) \text{ cal gm}^{-1}$$

where T is the temperature in degrees Celsius.

The large heat capacity of water and the latent heats of fusion and evaporation are significant in considering temperature changes. Much of the surplus heat of the tropical regions is used in evaporating water, the heat being released when the vapor changes back to the liquid state to form clouds. The annual temperature cycle of coastal areas and islands is modified because of the high heat capacities of the oceans (see Problem 6). Also, there is a lag in the temperature curve when it passes through the freezing point. During periods of falling temperature, the release of the latent heat of fusion provides an extra source of heat. When the temperature is rising, the heat required to melt snow or ice prevents a rapid warming. These latter effects are more significant near bodies of water than over land, and more significant in moist areas than in deserts.

Based on the above, one can determine the heat Δh required to raise the temperature of a mass m an amount ΔT by the equation[1]

$$\Delta h = mc \, \Delta T$$

Heat, a form of energy, can be obtained from other energy sources. The kinetic energy of a moving body, when stopped by friction, is transformed into sensible heat, warming the body. Part of the work done in compressing a gas, as into a bicycle tire, appears as sensible heat. The conversion factor between calories and ergs, the centimeter-gram-second (abbreviated cgs) unit of energy, is

$$1 \text{ cal} = 4.187 \times 10^7 \text{ ergs}$$

[1] In mathematics the symbol Δ is often used to measure an increment of a particular variable. Thus, if the temperature rises from 12°C to 15°C, $\Delta T = 3$°C. Unlike many algebraic formulas, the expression ΔT cannot be considered as a product but as representing a single quantity.

An *erg* is the energy expended when a force of one dyne acts for one centimeter. A *dyne* is the force that gives a mass of one gram an acceleration of one centimeter per second per second (1 cm sec⁻²).

2.9 *Charles' Law*

Most bodies, when heated, expand an amount proportional to the rise in temperature. The rate of expansion for solids and liquids varies from substance to substance. For gases there is a general rule known as *Charles' law* named for the scientist who formulated it. The law states that the expansion of a gas varies directly as the change in temperature, pressure being kept constant, and the constant of proportionality is 1/273 of the volume at 0°C. In algebraic form, this can be stated as

$$\Delta V = \frac{1}{273} V_0 \Delta T \tag{1}$$

where V_0 is the volume at 0°C.

Let the initial temperature be 0°C. Adding V_0 to both sides, we obtain

$$V_0 + \Delta V = \frac{1}{273} V_0 (273 + \Delta T) \tag{2}$$

If we now change our scale of temperature by adding 273° to the Celsius temperature (that is, using the Kelvin temperature) then,

$$V = \frac{1}{273} V_0 T \tag{3}$$

where now V is the volume at temperature T in °K. If V_1 and V_2 are, respectively, the volumes at T_1 and T_2, Equation 3 may be changed to

$$\frac{V_2}{V_1} = \frac{T_2}{T_1} \tag{4}$$

which is another method of stating Charles' law.

EXAMPLE: A balloon at 17°C has a radius of 0.5 m. What will be the radius when it has been cooled to −10°C?

In terms of Equation 4

$$T_1 = 17°C = 290°K \qquad V_1 = \frac{4}{3}\pi(0.5)^3 \text{ m}^3$$

$$T_2 = -10°C = 263°K \qquad V_2 = \frac{4}{3}\pi r_2^3$$

where r_2 is the final radius.

$$\frac{\frac{4}{3}\pi r_2{}^3}{\frac{4}{3}\pi(0.5)^3} = \frac{263}{290}$$

$$r_2 = 0.5\left(\frac{263}{293}\right)^{\frac{1}{3}}$$

$$= 0.48 \text{ m}$$

PROBLEMS AND EXERCISES

1. With a thermometer, check the variations of temperature in your vicinity. Of particular interest are the variations on a warm summer afternoon with variable cloudiness and light winds. If possible, the thermometer should be sheltered from the sun and be moved rapidly through the air just before reading. The reading should be to the nearest 0.1°F if possible and definitely to the nearest 0.5°F. The suitable places to be checked would be over a grass plot, in the shade of a building, over asphalt or concrete, in a wooded area, and near a lake or river, etc.

A second time when temperatures should be compared is a clear calm evening after sundown or, better still, about sunrise. Points to be checked include a valley or at the bottom of a hill, the top of the hill, a forest area, and an adjacent clear area.

2. Obtain, if possible, monthly temperatures for a winter month and a summer month for a small area with considerable variation in height, for instance, western Colorado. Plot the mean monthly temperatures against height. From your chart, estimate the rate of change of temperature with height. Is there any difference between the two seasons? (National weather services usually issue monthly climatological data from which the information can be obtained.)

3. Examine the temperatures on a weather map. Notice the general drop in temperature with increasing latitude. To what extent do temperatures at constant latitude differ? Can you attribute some of the differences to (a) the presence of water bodies and (b) to the altitude of the stations (see Exercise 2)? Are there other variations not explained by water bodies or altitude? (Isotherms drawn on the map will help bring out the variations in temperature.)

4. A balloon is filled with gas at a temperature of 86°F to make a sphere with a diameter of 1.5 m. What will be the diameter if it cools to —4°F?

5. Keep a record of temperatures that includes six or more readings at regular times during the day for a period of 5 days. Compare the diurnal ranges of temperature during the period. What explanations can you give for the variation?

6. From monthly mean temperatures for a large area, determine the annual ranges of temperatures. Your area should include coastal and inland stations. What are the causes for the variations in this variable?

7. A layer of snow 5 cm thick, density 0.24 gm cm^{-3}, warmed 10°C during a period of 4 hrs from 0930 h until 1330 h. Determine the rate of absorption of heat per unit area by the snow surface. Assume no melting or evaporation.

8. During a morning, a layer of 20 cm of sandy soil at the surface of the earth warmed on the average 6°C. If the specific heat of the sand is 0.2 cal gm^{-1} deg^{-1}, and the density 1.4 gm cm^{-3}, determine the amount of heat absorbed per unit area.

9. The air in a room 6 m $\times$ 5 m $\times$ 3 m has a density of 1.2×10^{-3} gm cm^{-3} and a temperature of 20°C. A mass of 5 kg of ice, temperature 0°C, is brought into the room. If heat from the air is used to melt the ice, what is the temperature of the air when the ice has turned to liquid? (In an actual situation, the walls of the room will supply heat so that the drop in temperature will be less than the value computed.)

10. Heat from 1 kg of air at constant pressure and at a temperature of 20°C is used to evaporate 1 gm of water. By how much is the air cooled?

CHAPTER 3

PRESSURE IN THE ATMOSPHERE

3.1 *Definition of Pressure*

An important measurement for meteorologists is the height of a column of mercury. This seems odd, for there seems to be little connection between this height and its changes and the current and future weather. Nevertheless a relationship exists, as Torricelli discovered. In 1643, he made the first barometer by filling a tube about one meter in length with mercury and inverting it into a basin of mercury. The mercury dropped from the top of the tube to a height of approximately 76 cm (30 in.). Soon after his invention, he noticed that the height did not remain constant, but was high during fine weather and low during rainy weather. The causes for this relationship have been the subject of much study. They are discussed in Chapters 6, 7, and elsewhere in the text.

A force may be exerted on a solid and, because of its rigidity, the solid will transmit the force to another body. With fluids this is not true. If a force is exerted on a fluid that is free to flow, it will move aside rather than oppose the force. If the liquid is contained in a vessel, a force exerted as with a piston is transmitted and acts equally on all the sides of the vessel. The force on a unit area is called the *pressure*. There are, of course, two forces on the area—the pressure by the external force and the counter-acting pressure by the fluid. This is the *static pressure* with which meteorologists are concerned. A fluid in motion exerts a force called a *kinetic force* against an obstruction. Moving air drives sailing ships and blows down buildings because of the pressures exerted but, in general, this kinetic pressure has little importance in the study of the weather. It is used at times to determine the velocity of the wind.

Consider a level surface S (Figure 3.1) in a vertical column of water. The liquid above S is attracted by gravity to the earth and, hence, there is a pressure F_1 from above because of the weight of the water. Consequently there must be an upward pressure F_2 on S to counteract the force of gravity. The pressure F_1 is equal to the weight of a unit column of the fluid above S and, therefore, increases with depth. The force of a jet of water from the base of a dam is much greater than aware of this increasing pressure with depth and must prepare for it from a jet half way up because of the increase in pressure. Divers are when they plan to go far below the surface. Air too exerts a static pressure because of the weight of the air above the level of measurement. This pressure decreases as one rises in climbing a mountain.

Pressure can be measured in grams weight per square centimeter or in pounds per square inch. In the cgs system, the fundamental unit of force is the dyne, and the corresponding unit of pressure is dynes cm^{-2}. A pressure of one gram weight per square centimeter equals approximately 981 dynes cm^{-2}, and would give to a gram mass an acceleration of 981 cm sec^{-2}. This acceleration due to gravity varies with latitude, the sea-level value increasing from 978.0 cm sec^{-2} at the equator to 983.2 at the poles. The value of 980.616 cm sec^{-2} is taken as a standard for many meteorological measurements.

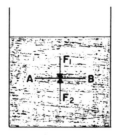

FIGURE 3.1 Pressures on a surface immersed in a liquid.

3.2 *Measurement of Pressure*

Torricelli's measurement of pressure was based on the law that the pressure at a given level within the fluid is constant (see Figure 3.2). When the barometer is at sea level, the mercury in the tube falls to approximately 76 cm above the level of the mercury in the basin. By the law of pressure, the mass of air above a unit area at A must equal the mass of mercury above a unit area at B. This latter is

$$13.6 \text{ gm cm}^{-3} \times 76 \text{ cm} \times 1 \text{ cm}^2 = 1034 \text{ gm}$$

where 13.6 gm cm^{-3} is the density of the mercury. The force acting on the area is 1034 gm $\times$ 981 cm sec^{-2} = 1.014 $\times$ 10^6 dynes. Therefore, the pressure is 1.014 $\times$ 10^6 dynes cm^{-2}, or in the foot-pound-second system of units 14.7 lb in^{-2}.

The unit of dynes cm^{-2} is very small and, therefore, other units are substituted for practical use. The *bar*, defined as 10^6 dynes cm^{-2}, is approximately the force exerted by the atmosphere at sea level, and is sometimes used. More common is the *millibar* (mb), which is 10^{-3} bars. Sea-level pressure in the standard atmosphere is defined as 1013.25 mb. Sometimes people speak of the pressure in terms of the height of the mercury column. In these units, the pressure of the standard atmosphere is 760 mm or 29.92 in. mercury. These values permit a comparison of values in the three different units. Approximate relationships are

$$3 \text{ mm mercury} = 4 \text{ mb}$$
$$3 \text{ in. mercury} = 100 \text{ mb}$$

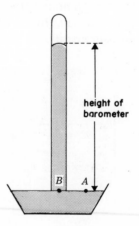

FIGURE 3.2 Simple barometer.

FIGURE 3.3 Barograph with cutaway view showing chamber. (Courtesy, Science Associates.)

The actual heights of mercury read on a barometer cannot be translated into millibars directly by the above relationships. They must first be adjusted for the value of the acceleration due to gravity and the temperature of the column of mercury, which changes its specific gravity.

The principles underlying Torricelli's barometer are still the fundamental principles by which pressure is measured accurately. The refinements for our modern barometers are to ensure that the temperature of the mercury is known and that the height of the column of mercury is read with extreme precision. The final result must give the pressure to the nearest 0.1 mb. This means that the height of the mercury column must be read to the nearest 0.003 in. or 0.1 mm.

The mercury barometer is, by nature, difficult to move from one location to another, and cannot stand rough usage. A substitute barometer is the *aneroid* (meaning without liquid) *barometer*. A partially evacuated metal box with corrugated sides (*A* in Figure 3.3) expands or contracts as the pressure of the atmosphere against it falls or rises. By calibrating these changes with the aid of a mercury barometer, it is possible to use them to determine the pressure. When a pointer B is connected to the metal box, the changes in pressure may be read from the changes in the pointer or may be recorded by means of a pen on a revolving drum C. This instrument is called a *barograph*. Adaptations of the aneroid barometer are also used to indicate pressure in household barometers, on ships, or on aircraft. These barometers, for a number of reasons, do not record air pressure as accurately as does a mercury barometer. Also, the sources of error gradually alter with time. Therefore, an aneroid barometer cannot be employed as a precision instrument, but its portability makes it very useful.

3.3 *The Gas Laws*

In Section 2.9, Charles' law for gases was translated into the mathematical equation

$$V = K_1 T \tag{1}$$

where T is the absolute temperature and V the volume. This is true provided that the pressure on the gas does not change. Another fundamental relationship for gases is Boyle's law, which states: when the temperature is kept constant, the volume of a gas multiplied by its pressure is constant, that is,

$$pV = K_2 \tag{2}$$

where p is the pressure and V the volume. Dividing by the mass m of the gas, which of course is constant, one obtains

$$pv = \frac{K_2}{m}$$

where v is the volume per unit mass, or *specific volume*. In meteorology, the *density* ρ, that is, the mass per unit volume, is more generally used than the specific volume. These two quantities are reciprocals, that is,

$$\rho = \frac{1}{v}$$

Therefore,

$$p = \frac{K_2}{m} \rho \tag{3}$$

which is another mathematical form for Boyle's law.

Equations 1 and 3 can be combined (see Exercise 12) into the form

$$p = \rho RT \quad \text{or} \quad pv = RT \tag{4}$$

where R is known as the gas constant. For air, the value of R in the cgs units has been determined to be 2.87×10^6 ergs gm^{-1} deg^{-1} or cm^2 sec^{-2} deg^{-1}.

Equation 4 can be used in a number of ways. For instance, let us determine the density of air under standard pressure (1013 mb) and at a temperature of 68°F. Before we use Equation 4, it is necessary to convert pressure to dynes cm^{-2} by multiplying by 10^3. Also, the temperature must be converted to degrees Kelvin

$$68°F = 20°C = 293°K$$

Therefore,

$$\rho = \frac{1013 \times 10^3 \text{ gm cm}^{-1} \text{ sec}^{-2}}{2.87 \times 10^6 \text{ cm}^2 \text{ sec}^{-2} \text{ deg}^{-1} \times 293 \text{ deg}}$$

or

$$\rho = 1.205 \times 10^{-3} \text{ gm cm}^{-3}$$

This amounts to 1.205 kg m^{-3}, or approximately $\frac{3}{4}$ lb in 10 cu ft. Under these circumstances, the mass of air in a room 4 m $\times$ 3 m $\times$ 3 m is 36 m^3 $\times$ 1.2 kg m^{-3} = 43 kg or 96 lbs.

3.4 *The Hydrostatic Equation*

The barometer measures the pressure at the level of the mercury in the cistern. Because the pressure is determined by the weight of the air above the point of interest, the pressure changes with height by an amount equal to the weight in a column of air with unit cross section between the two levels.

The volume of this column (see Figure 3.4) equals $1 \times 1 \times (z_2 - z_1)$, where z_1 and z_2 are the heights of the two levels. If ρ is the density, then the mass of the column of air is $\rho (z_2 - z_1)$. By using the prefix Δ, we then may write

$$\Delta p = -g\rho \, \Delta z \qquad (5)$$

The negative sign must be inserted since the pressure decreases as the height z increases. This formula is known as the *hydrostatic equation*.

Equation 5 can be used in a number of problems as, for example, to find the change in pressure from the bottom to the top of a 100-foot high building. By working in cgs units, and by using the value of the density found in the preceding section,

$$\Delta p = -981 \text{ cm sec}^{-2} \times 1.205 \times 10^{-3} \text{ gm cm}^{-3} \times 100 \text{ ft} \times 30.5 \text{ cm ft}^{-1}$$
$$= -3.605 \times 10^3 \text{ dynes cm}^{-2}$$
$$= -3.6 \text{ mb}$$

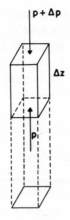

FIGURE 3.4 Change of pressure with height.

In making this computation, we have assumed that the density has remained constant, whereas it has actually decreased with pressure. The answer is sufficiently accurate for the purpose of the problem. For thicker layers, it is necessary to determine a mean pressure and temperature from which to compute the density. Students of calculus will recognize that Equation 5 may be combined with Equation 4 and, assuming a mean temperature for the layer T, transformed into

$$\log \frac{p_0}{p} = 0.434 \, \frac{g}{RT} \, (z - z_0)$$

$$= 0.434 \, \frac{g}{RT} \, \Delta z \tag{6}$$

where p_0 is the pressure at z_0 and p the pressure at z. The use of Equation 6 permits the calculation to be made with an assumed value of the mean temperature only. For the purpose of this text either Equation 5 or 6 gives sufficiently accurate answers.

3.5 Pressure and Altimetry

By means of the relationship given in Equations 5 and 6, it is possible to determine height changes by observing pressure changes. Aircraft pilots do this by means of an *altimeter*. This instrument is basically an aneroid barometer, but with the pressure scale replaced by a height scale. The relationship between height and pressure given in the hydrostatic equation depends on density, which is a function of both pressure and temperature. Thus, with any vertical temperature profile, there is a one-to-one relationship between height and pressure fall. By international agreement, the scale of an altimeter is based on the U.S. Standard Atmosphere, described in Section 2.7. Table 2.1 gives the heights corresponding to different pressures. An error in the reading may arise either because the current sea-level pressure differs from the standard value of 1013 mb, or because the actual temperatures differ from the ones given in Table 2.1.

A plane traveling according to the altimeter at constant altitude is in reality traveling along a surface where the pressure is constant but not necessarily at a constant height. To the extent that the atmosphere below the plane differs from the standard atmosphere, the recorded altitude differs from the true altitude of the plane. This departure of the altimeter reading from the true altitude is significant in two situations. The pilot needs to know his true altitude, both as he lands at an airfield and, also, as he flies near mountain peaks. It is also necessary that the vertical separation between aircraft is sufficient to ensure safety.

The error of an altimeter can be significant when the aircraft is flying low over the ocean or over areas where the altitude is near sea level if the pressure at sea level differs from the standard 1013 mb. When the sea-level pressure is high, the altimeter reading is too low, and vice versa. Normally, over ocean areas, the altimeter setting is kept at 1013 mb, and the pilot is alert to the error in height if he is flying low over the sea. For flying over land near sea level, he adjusts for the pressure by changing the dial of the altimeter so that it will read zero altitude when he is at sea level. The sea-level pressure reported to the aircraft is called the *altimeter setting*. The altimeter height will be close to the true height as long as he is close to sea level.

To consider the problem of landing at an airport A situated above sea level, see Figure 3.5. A pilot flying above the sea will set his altimeter on the basis of the pressure at B (that is, at sea level). His altimeter will read correctly if the temperature profile below him is the same as the standard atmosphere. Thus, when flying where the pressure is 950 mb, he will read an altitude of 540 m (see solid lines). If the air below him is warmer, the altimeter will read too low.

The effect of temperature on indicated altitude is best seen in Equation 6. For instance, if the surface temperature is 30°C and the lapse rate is 6.5°C km^{-1}, the pilot will be at 568 m when the pressure is 950 mb, but his instrument will still read 540 m. In cold air, the altimeter will read too high. If the pilot approaches an airport A that is close to the coast but has an elevation of 1 km, the air being under standard conditions, the pressure at the airport and at airport level is 899 mb. If again the air is 30°C at sea level and the lapse rate 6.5°C km^{-1},

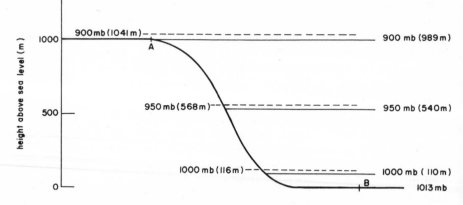

FIGURE 3.5 Constant pressure surfaces for an aircraft approaching an airport 1 km above sea level. Solid lines are based on the standard atmosphere; dashed lines on a sea-level temperature of 30°C and a lapse rate of 6.5°C km⁻¹.

the aircraft flying at an indicated altitude of 1 km will actually be at 1053 m and, hence, will be too high to land.

This problem is solved by adjusting the altimeter setting. Under standard conditions, the pressure difference between sea level and 1 km is 114 mb. If the pilot then sets his altimeter at the value given by the station pressure in millibars plus 114, the instrument will record his height as 1 km when he lands. In North America, the altimeter settings are given in inches of mercury, but the principle is the same. With the same setting, the altimeter is also approximately correct in the vicinity of the airport and, thus, the pilot knows at approximately what indicated altitude he must fly to clear hilltops in the vicinity.

EXAMPLE: Given that at a station 500 m above sea level the pressure is 960 mb. Determine the altimeter setting.

With a sea-level temperature of 15°C and a lapse rate of the column 6.5°C km⁻¹ the mean temperature of the fictitious layer below the airport is 13.4°C = 286 °K. Assuming a mean pressure of 980 mb, and by using Equation 5,

$$\Delta p = -\frac{981 \times 980 \times 10^3}{2.87 \times 10^6 \times 286} (-5 \times 10^4) \text{ dynes cm}^{-2}$$
$$= 58.6 \text{ mb}$$

The altimeter setting is 1018.6 mb. Equation 6 gives the setting to be 1019.0 mb.

The problem of separation of aircraft is more easily solved. If two aircraft, one above the other, are using the same altimeter setting, the recorded altitudes may both be in error. But the errors will be in the same direction since they arise from the same temperature distribution, with the result that the difference in the altimeter readings will give approximately the difference in the aircraft heights. Even when the heights are seriously in error, the aircraft will be flying with safety. International agreement specifies that over ocean areas the altimeter be set at a sea-level pressure of 1013 mb, thus ignoring the variations of pressure that are observed. Over the land areas, the altimeter setting is obtained by radio from the nearest airport. In this manner, vertical separation of planes flying at different indicated levels is kept constant.

3.6 Sea-level Pressure

The pressure gradients in the vertical, discussed in the preceding section, are large. The rate of change of pressure between two points at the same height is much smaller. Thus, a change in the horizontal of 8 mb in 100 km is very large, although in the vertical the same change will occur in 100 m or less. Nevertheless, as will be shown in Chapter 7, outside the tropical regions this *horizontal pressure gradient* is the chief factor in determining the wind and the weather. Therefore, it becomes important that these gradients be determined accurately.

If two stations are at the same level, it is not difficult to determine the horizontal pressure gradient. When the two points, *A* and *B* of Figure 3.6, are at different levels, the information desired is the pressure difference in the free atmosphere, that is, between *B* and *C*, a point above *A* and at the level of *B*. The pressure at *C* can be determined

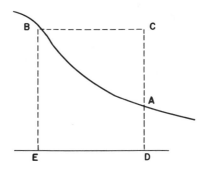

FIGURE 3.6 Determination of the horizontal pressure gradient.

by subtracting from the pressure at A the pressure change between A and C as given by the hydrostatic equation, a pressure difference that changes with changing temperatures.

For an area-wide comparison, it is desirable to establish a base level, usually sea level. In Figure 3.6 this is given by the line DE, where D and E, respectively, are below A and B. It is necessary to compute "sea-level pressures" for D and E so that the difference between them is approximately that between pressures at B and C. This is done by assuming that a column of air exists from B to E and from A to D with temperatures based on the surface temperatures. For D, the pressure difference is computed for the column AD, and the result added to the surface pressure at A. A similar computation by the observer at B gives the sea-level pressure at E.

The hydrostatic equation requires that a temperature be assigned to these fictitious columns of air from the surface to sea level. In using the equation, the surface temperature is usually taken as the average between the current temperature and the one occurring 12 hours before, thus eliminating the diurnal variation of temperature. The lapse rate is taken as 6.5°C km⁻¹. Nevertheless, as can be observed, the adjustment to sea level can be based on different assumptions by different people. There can be no unique "sea-level pressure" for a station above sea level. The only requirement in determining a value is that the method be consistent within a region so that the horizontal pressure gradient between D and E of Figure 3.6 gives a close approximation to the actual pressure gradient between C and B. There will be times, such as when A is in warm air on one side of a mountain ridge and B in cold air on the other side, when the desired result will not be obtained. When, as in Antarctica, and Greenland, and during the winter in Siberia, Alaska and Yukon, surface temperatures are low, the adjustments to sea level are high. The extremely high "sea-level pressures" may not then represent the pressure distribution in the free air correctly.

EXAMPLE: Compute the sea-level pressure for the station in the example of Section 3.5. Assume a station temperature of −20°C, and a lapse rate in the layer below the station of 6.5°C km⁻¹.

The mean temperature for the column is $-20 + 1.6 = -18.4°C = 255°K$. An approximate mean pressure is 980 mb. Then

$$\Delta p = \frac{-981 \times 980}{2.87 \times 10^6 \times 255}\,(-5 \times 10^4) \text{ mb}$$
$$= 65.7 \text{ mb}$$

Sea-level pressure, according to this calculation, is 1025.7 mb.

3.7 *Maps of Sea-level Pressure*

As indicated in Section 3.6, pressure and its horizontal distribution are fundamental in understanding the weather. A complete report on the weather includes the sea-level pressure for the station. The barometer is read, and the value corrected for the temperature of the mercury and the errors of the barometer to give the pressure at the station level. From this is obtained the sea-level pressure.

As any barograph record shows, the pressure of the atmosphere is continually changing. Normally the changes are slow, as shown in Figure 3.7, which gives the trace of the barograph at Edmonton, Alberta for the week of 9 to 15 March 1964. More rapid changes occur when a hurricane or a tornado approaches. Normally sea-level pressures remain in the range between 980 and 1020 mb. It is rare that the sea-level pressure departs from the range 960 to 1050 mb. The highest recorded sea-level pressure for the world is 1084 mb at Agata, Siberia on 31 December 1968; in North America, 1065 at Dundas, north of Thule, Greenland on 18 January 1958. The lowest recorded pressures have been at the centers of hurricanes. The record low for the world is 877 mb in a Pacific typhoon on 24 September 1958; for the United States, 892 mb at Long Key, Florida, 2 September 1935. It is probable that much lower pressures than these occur at the centers of some tornadoes, but the information is lacking.

Values of the sea-level pressures are observed at the synoptic hours and then collected at weather offices across the country. The values plotted on maps permit meteorologists to draw *isobars,* or lines of equal pressure. Figure 3.8 gives the isobars for eastern North America for 1200 GMT 8 January 1969, based on the sea-level pressures at that time. Figure 2.9 gives the temperatures and winds for the same time. As is usual in weather maps, the value of the pressure is plotted at the upper right-hand side of the station circle. The value plotted is the tens, units, and tenths digits of the pressure with the decimal point omitted. Because the value usually lies between 950 and 1050 mb, the hundreds digit can be supplied without ambiguity. The isobars in Figure 3.8 are drawn at intervals of 4 mb, a common but not universal interval. The *tendency,* that is the change in the pressure in the preceding three hours, is plotted below the pressure. A mark after the tendency represents, pictorially, the manner in which the pressure changed.

The map shows features common to many weather maps. Over the southeastern United States there was a region of high pressure, sometimes called a *high* or *anticyclone.* Another high lay off the map, in

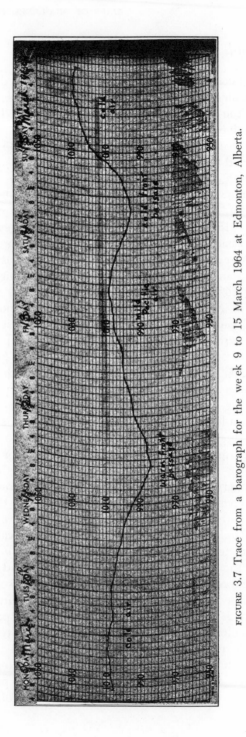

FIGURE 3.7 Trace from a barograph for the week 9 to 15 March 1964 at Edmonton, Alberta.

the Canadian Northwest Territories. A *ridge* through Michigan connected the two anticyclones. A center of low pressure, called a *low* or *cyclone*, was present over western Kansas and eastern Colorado, with a *trough* northward into Minnesota. The term cyclone is sometimes used by the public to designate a tornado. To the meteorologist, a tornado is only a special kind of cyclone, although a very dramatic kind. The map shows that a second low lay over the Maritime Provinces of Canada. The point of lowest pressure on the line between two highs and the point of highest pressure on the line between two lows is called a *col.* One was present over Michigan at the time of the map.

When weather maps are completed, the pattern shows that the pressure change from place to place is sometimes gradual and sometimes rapid, but never abrupt. They are similar in some respects to topographic maps with their contours, but an isobaric map would never have a place which would correspond to a perpendicular cliff. Isobars on a weather map cannot touch.

In equatorial areas, the pressure varies little from day to day because here there are no migrating cyclones and anticyclones such as are found in latitudes greater than 30°. Barograph traces for the tropics show that a pressure rhythm exists, governed by the time of day. Pressures tend to reach maximum values at 1000 and 2200 h, local time, and minimum values at 0400 and 1600 h. The difference between extremes is only 3 mb, a small variation compared with the much greater swings in temperate and polar regions. By careful analysis it is possible to show that a smaller diurnal variation is present in these latter regions, but is masked on individual days by the much greater swings caused by moving pressure centers.

3.8 *Upper-level Pressure Maps*

Sea-level pressure maps are drawn by analyzing the distribution of pressure using isobars. Initially, the maps above the surface of the earth were drawn for constant levels, that is, 5000 ft, 10,000 ft, etc. This system was changed to simplify the analysis. Now upper-air stations report the heights above sea level where the pressure has dropped to specified values. These pressure values are at present 1000, 850, 700, 500 mb, etc. For any one pressure level, these heights are plotted on a map, and lines of constant height are drawn. The contours of this invisible surface may be used by the meteorologists as a guide to wind and weather changes at the earth's surface. Because the sea-level pressure is related to the height of a standard pressure surface, the surface and upper-air charts are closely interrelated. Figure 3.9 gives the map

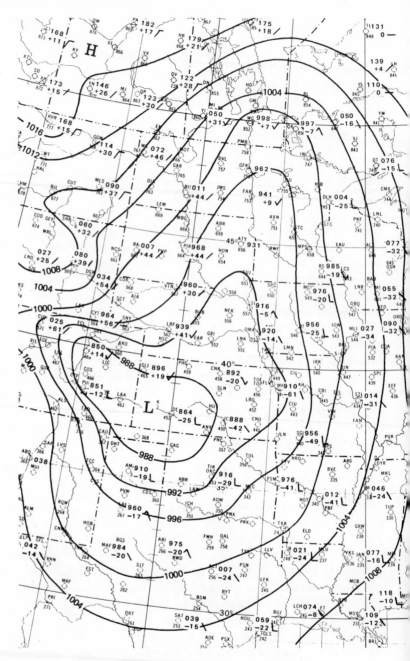

FIGURE 3.8 Isobars for eastern North America, 1200 GMT 8 January 1969. Pressures and tenths digits only. Below the pressure is the change in pressure in tenths of change. This map should be compared with Figure 2.9.

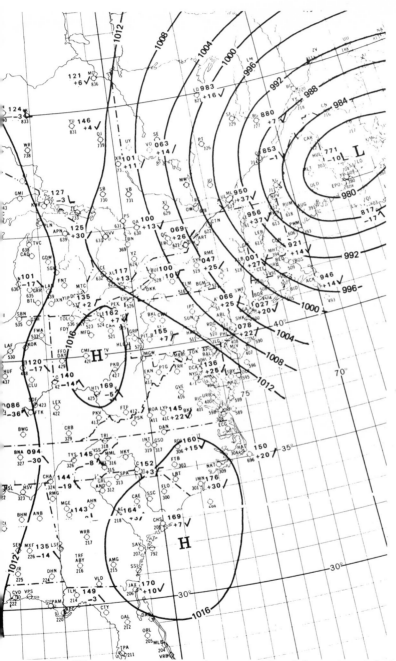

are in millibars and are given to the upper right of the station circle in tens, units,
a millibar during the preceding three hours, with a line illustrating the manner of

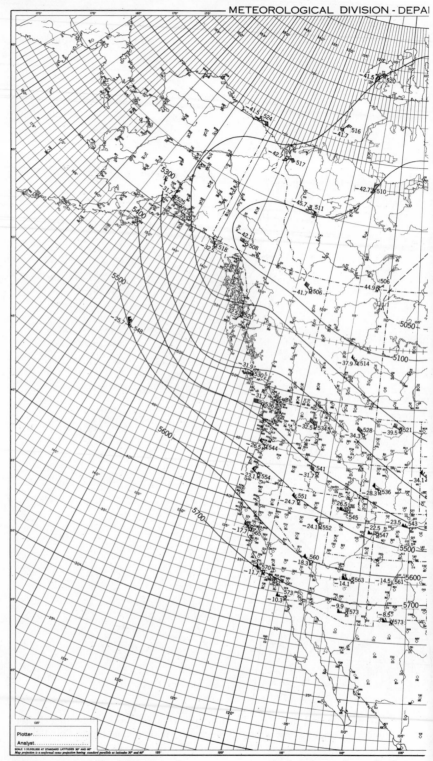

FIGURE 3.9 Contours of the 500-mb surface in meters, with the observed winds, for (26 m sec⁻¹) and a full barb 10 knots (5 m sec⁻¹).

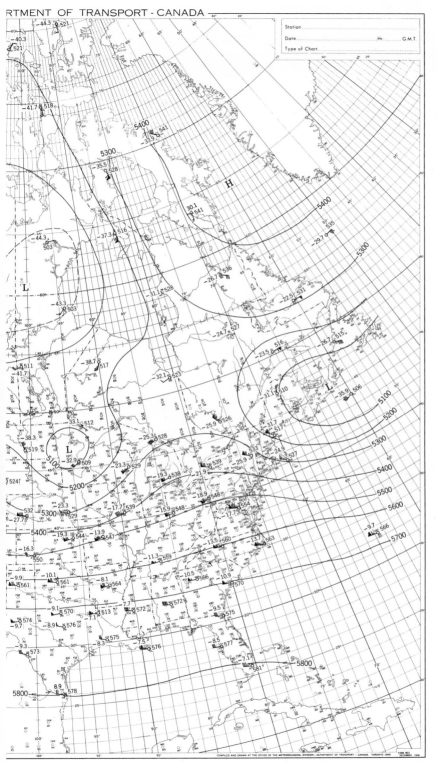

Station ....

Date .... Hr .... G.M.T.

Type of Chart ....

1200 GMT 8 January 1969. A flag on the wind shaft indicates a wind of 50 knots

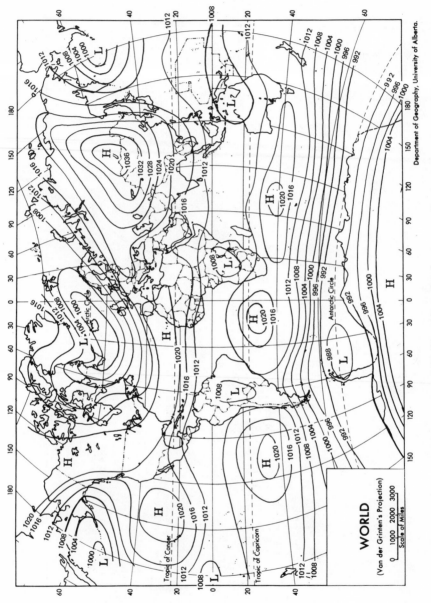

FIGURE 3.10 Normal sea-level pressure, January.

Department of Geography, University of Alberta.

WORLD
(Van der Grinten's Projection)

0   1000   2000   3000
Scale of Miles

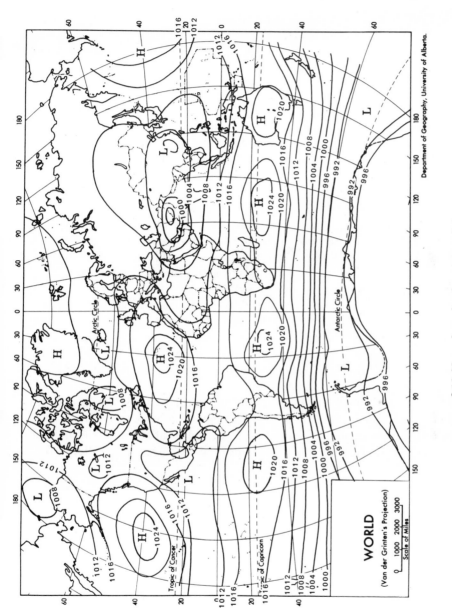

FIGURE 3.11 Normal sea-level pressure, July.

Department of Geography, University of Alberta.

61

of the 500-mb surface for 1200 h GMT 8 January 1969, that is, for the same time as Figures 2.9 and 3.8.

3.9 *Maps of Mean Pressure*

Synoptic charts such as Figure 3.8, giving the distribution of pressure over the area for a specific time, are useful in explaining the weather for that time. For this reason they form the basic tool of the weather forecaster. When average values of pressure for the stations of an area are calculated and maps are drawn from these values, much but not all the variation found on individual charts disappears.

Mean pressures are usually determined for individual months. By averaging the means for a period of years, one may obtain the normal sea-level pressure for a station. From the normals for an area, the normal sea-level distribution may be determined. The world normal distributions for the months of January and July are given in Figures 3.10 and 3.11, respectively. They show that, in certain areas of the earth, pressures tend to be high. Examples are found over the ocean at 30–35° latitude. Other regions, for example near Iceland and along the equator, tend to have low pressure. In Asia, the pressures tend to vary considerably with the season. Because these maps provide a background for the study of the climates of the earth, reference will be made to them in other parts of the text.

The departures of pressures from the normal values for a specific period may also be analyzed on a geographical basis, and conclusions can be made regarding the departures of the weather from the normal weather for that time of year. An example is given in Section 9.3.

PROBLEMS AND EXERCISES

Note. In using the hydrostatic equation, mean pressures and temperatures must at times be assumed. Answers will then not necessarily correspond with the ones given on page 309, but the differences should be small.

1. As a hurricane approached, the pressure dropped 75 mb. Compute the decrease in kilograms in the mass of air over one square kilometer.

2. A rigid vessel contains dry air at 25°C and 1000 mb pressure. Compute the pressure if the temperature is raised to 100°C. If the mass of the air was 3 kg, determine the heat required for the process.

3. In Section 3.5, the drop in pressure from sea level to 1 km is given as 114 mb, based on the standard atmosphere. By making use of the hydrostatic equation and suitable assumptions, check this value.

4. As a weather system approached, the pressure dropped from 1032 mb to 982 mb. If a house, 15 m $\times$ 12 m $\times$ 10 m, is airtight, what is the resultant force acting on the vertical walls of the house? If the house is not airtight (so that pressures on the inside and outside of the house are equal), compute the loss of mass of the air in the house.

5. Find the difference in height between 500 and 300 mb when the mean temperature is $-38°C$. By what amount must the temperature rise to increase the difference by 50 m?

6. At two points, A and B, both at 2000 m above sea level, the pressure is 800 mb. At A the temperature is $2°C$ and the lapse rate below A is $5°C$ km^{-1}. At B the temperature is $-3°C$, and the air is isothermal below B. What is the difference in sea-level pressures below the two points?

7. In Exercise 6, compute the altimeter settings for A and B.

8. An airplane flying at approximately 1 km above mean sea level records a pressure of 920 mb. Sea-level pressure is 1010 mb and the air column has the temperature of the standard atmosphere. Compute the height of the plane.

9. If in Exercise 8, the mean temperature of the air below the airplane is $-3°C$, what is the error in the altimeter reading?

10. A traveler carries with him an aneroid barometer which gives measurements of pressure in inches of mercury. Determine an approximate relationship with which he can convert a pressure change of 0.1 in. to a change in altitude.

11. In the example in Section 3.5, the mean pressure was taken as 980 mb. What would have been the altimeter setting if the value had been taken as 990? What change would there have been if the mean temperature had been taken as $285°K$ instead of $286°K$?

12. A mass of gas has initial conditions of p_0, v_0, and T_0, intermediate conditions of p_1, v', and T_0, and final conditions of p_1, v_1, and T_1. Use Equations 1 and 2 to determine a relationship among p_0, v_0, and T_0 and p_1, v_0, and T_1. From the relationship obtain Equation 4.

13. A 50-m high building is kept at $20°C$ when the outside temperature is $-20°C$. If pressures inside and outside at the top of the building are equal, compare the pressures at the bottom of the building. Consider the significance of your result in explaining winds in and out of doors and windows.

CHAPTER 4

MOISTURE
IN THE
ATMOSPHERE

4.1 *Physical Properties of Water*

Water, with its peculiar characteristics, and its changes from one state to another at normal atmospheric temperatures, is the subject of considerable study in many of the earth sciences, especially in meteorology. In fact, many changes in weather are merely changes in the state of the ever-present water.

Within the normal range of temperatures, water may exist in any of the three states, solid, liquid, and gaseous. Some of its physical properties are of special significance for weather processes. Normally, at temperatures below 0°C the substance exists as a solid (ice) with a density of 0.92 gm cm^{-3}. Its specific heat at 0°C is 0.50 cal gm^{-1} deg^{-1}, but this decreases with decreasing temperature. When ice melts, it absorbs the latent heat of fusion of ice, 80 cal gm^{-1}.

Water as a liquid is found under normal atmospheric conditions between 0° and 100°C. With temperatures below freezing, even as low as —40°C, pure water may still exist in the atmosphere as small supercooled droplets, but they will freeze immediately on hitting a solid object such as an airplane or a car windshield. The density of water changes slightly with temperature, with maximum density at 4°C. This characteristic permits the bottoms of lakes and rivers to remain above freezing when the surface water falls to 0°C. The specific heat is approximately 1 cal gm⁻¹ deg⁻¹, varying slightly with temperature. To evaporate water, at constant pressure, heat to the amount of (597 − 0.56 T) cal gm⁻¹ is required, where T is the temperature in degrees Celsius.

Water vapor, an invisible gas, may exist at all normal temperatures. The specific heat is, for constant pressure, 0.44 cal gm⁻¹ deg⁻¹, and for constant volume, 0.33. Its density varies over wide limits, but there is an upper limit, dependent on temperature, beyond which the density does not normally rise.

4.2 Measures of Water Vapor

Consider an enclosed vessel that has a small amount of liquid water at the bottom (Figure 4.1). Some of the water molecules, moving more rapidly than their fellows, escape into the space above the liquid. The rate of escape depends on the velocities of the molecules which, in turn, depend on the temperature of the liquid. As more molecules are present in the space above the surface, some of them collide with the surface of the water and become absorbed once again. For any temperature, there is an equilibrium condition when the two flows across the water surface are equal and, therefore, the amount of vapor remains constant. The space is then said to be *saturated*. Common usage speaks of the air being saturated, but the equilibrium condition is not dependent on the presence of other gases.

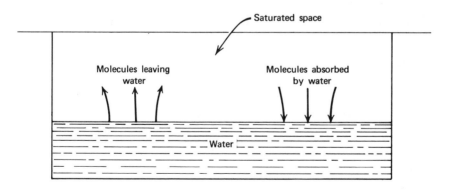

FIGURE 4.1 Development of saturated space.

Because the molecules of water vapor are in rapid motion, they exert a pressure by colliding with the sides of the container. This pressure is, by custom dating back to Dalton, denoted by e rather than p. It is related to the density of the vapor, ρ_w, by the gas law (Equation 4, Section 3.3).

$$e = \rho_w R_w T \tag{1}$$

In this equation, R_w, the gas constant for water vapor, is 4.62×10^6 cm^2 sec^{-2} deg^{-1}. From this relationship, we can determine the water vapor density from the vapor pressure, provided that the temperature is known. When the density has reached its maximum, the vapor pressure is the saturation vapor pressure for that particular temperature.

A simple measure of the amount of moisture in the air is the density of water vapor, in grams per cubic centimeter or in similar units. This is called the *vapor concentration* or the *absolute humidity*. The ratio of the actual density to the density required to saturate the space at the specific temperature is called the *relative humidity*. This is usually stated as a per cent. Notice that to find the moisture contained in the air we must know both relative humidity and temperature. Very dry air over a desert, relative humidity 5 per cent, can have more moisture than saturated very cold arctic air.

Another, and very common, measure of the moisture content is the *dew point*. This is the temperature at which the air, without changing its moisture content, would become saturated. According to the process defining saturation, the temperature uniquely determines the saturation vapor concentration. Therefore, from the dew point, we can determine the vapor present in the atmosphere and vice versa. A high relative humidity means that the difference between the temperature and the dew point is small, but the actual relationship is complicated. The term dew point has been derived, of course, from the fact that when air cools in the evening to the dew point temperature, dew usually begins to form.

Two measures of moisture relate the moisture in the air to the mass of the air itself. The *specific humidity* for moist air is given by the ratio

$$\text{specific humidity} = \frac{\text{mass of water vapor}}{\text{mass of dry air and water vapor}}$$

By Charles' and Boyle's laws, discussed in Sections 2.9 and 3.3, we learn that

$$\text{specific humidity} = 0.622 \frac{e}{p} \tag{2}$$

where p is the pressure of the air. The constant 0.622 is the ratio of the molecular weight of water to that of air.

The *mixing ratio* (w) in theory differs from the specific humidity by altering the denominator of the ratio to the mass of the air without the water vapor. Because the latter is generally less than 2 per cent of the total mass, and almost never more than 4 per cent, the difference between the two measures is minor. Notice that the mixing ratio and specific humidity do not change when the air mass expands or contracts. This is an example of a *conservative property*. A conservative property of the atmosphere is one that remains unchanged during certain specified processes. The mixing ratio and the specific humidity are conservative during movements of the air upward or downward, or with heating or cooling, provided that there is no evaporation or condensation. Conservative properties are very useful in meteorology because so many meteorological elements vary through air movements and other changes. The dew point is conserved during changes of temperature at constant pressure, again provided that no evaporation or condensation occurs. There is a slight change in the dew point when the pressure changes, but this is not significant at the surface of the earth (see Problem 13).

The mixing ratio required for saturation (w_s) depends on the air pressure. The values of the saturation mixing ratio at 1000 mb are given in Appendix 2. Figure 4.2a gives in graphical form the relationship between temperature and saturation mixing ratio. We observe that the values are small for temperatures below freezing, but that the curve rises rapidly with temperatures above 10°C. If the values of w_s are plotted on a logarithmic scale, as in Figure 4.2b, the relationship becomes approximately linear. A graph of this nature permits values to be read off quickly to two significant figures for all temperatures.

The relative humidity can be evaluated as the ratio e/e_s or equally well by w/w_s. By using Appendix 2 or Figure 4.2, we can begin with any two of the three quantities—temperature, dew point, and relative humidity—to determine the third. For example, consider air with a temperature 74°F (23.3°C), dew point 61°F (16.1°C). According to Appendix 2, the mixing ratio required for saturation at 74° is 18.4 gm kg^{-1}. The actual amount of moisture is sufficient to saturate the air at 61°, that is, 11.65 gm kg^{-1}. The relative humidity is then 11.65/18.4 or 63 per cent.

The preceding discussion has been based on temperatures above freezing. The situation is similar for temperatures below freezing, except that now the vapor may be in equilibrium with ice or frost. In general, amounts of water vapor at subfreezing temperatures are small, and also a small change in the absolute humidity alters the frost point markedly.

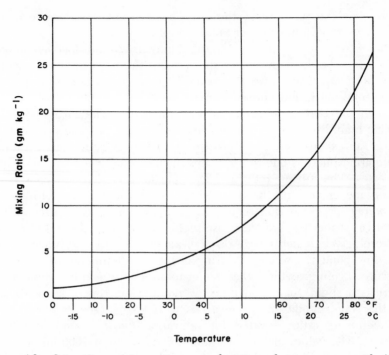

FIGURE 4.2a Saturation mixing ratio as a function of temperature with linear coordinates.

As pointed out in Section 4.1, it is possible to have supercooled water. When this happens, the vapor pressure for saturation over the super-cooled water is slightly higher than that over ice. Then it is necessary to specify that the relative humidity is based on saturation with respect to either ice or supercooled water. The difference will be found to have significance in Section 4.8, where the formation of rain is considered.

Of the various measures mentioned in this section, the one most commonly used by the layman is the relative humidity. Unfortunately, this lacks precision because it depends not only on the amount of moisture but also on the temperature. On the other hand, it does have meaning to the ordinary man for it gives a measure of the rate of evaporation. When it is high, evaporation is slow; when low, evaporation is rapid. The meteorologist prefers the measures of mixing ratio and dew point when he considers moisture in the atmosphere. The other measures are less commonly used, although the vapor pressure is used in meteorological computations.

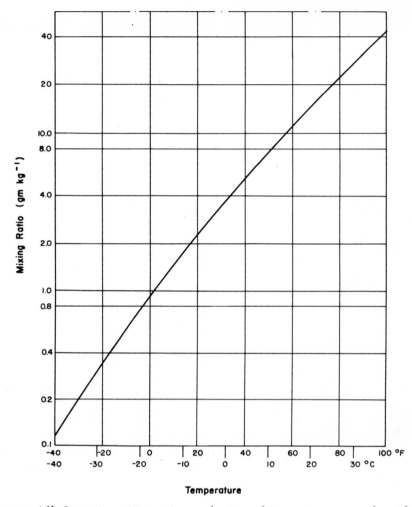

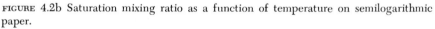

Temperature

FIGURE 4.2b Saturation mixing ratio as a function of temperature on semilogarithmic paper.

4.3 *Instruments for Measuring Water Vapor*

The definition of the dew point suggests a method of determining the moisture content of the air. Take a metal pitcher containing water into the area of interest. By adding snow or crushed ice, reduce the temperature of the metal until a film of water forms on the outside. Then cause the temperature to rise until the deposit evaporates. The average between the temperatures is the dew point. Appendix 2 gives the information necessary to determine the relative humidity.

The *dew-point hygrometer*, based on the foregoing principle, is used at times when close accuracy of the water-vapor content is required. A metal plate is cooled electrically or by other means to the point where moisture condenses on the plate. The time when this occurs can be detected by an electric eye that notes the reflection or nonreflection of a beam of light. This method is particularly valuable when the dew point increases rapidly with a small change in moisture.

A simpler instrument for measuring moisture is the *wet-bulb psychrometer* (Figure 4.3). It consists of two thermometers, one of which is kept moist by means of a wick leading out of a dish of water. The wet bulb usually reads lower than the dry bulb. To understand the principle of the wet-bulb temperature, consider 1 kg of air with a tem-

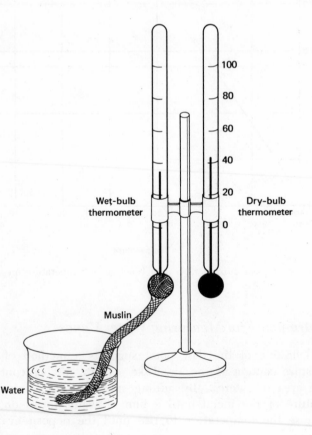

FIGURE 4.3 Diagram showing the principle of the wet-bulb thermometer.

perature of 30°C and mixing ratio of 7.76 gm kg⁻¹. According to Appendix 2, the dew point is 10°C and the relative humidity 28 per cent. Evaporation of 1 gm of water into 1 kg of air (see Problem 10, Chapter 2) cools the air approximately 2.5° to 27.5°C. At the same time, the extra moisture has increased the dew point to 11.8°C. This process can be continued until, at a temperature of 17.3°C, the air will be saturated, and no more water can be evaporated into it. This temperature is unique, given the initial temperature and moisture, and is called the isobaric *wet-bulb temperature*.

The process described occurs in the vicinity of a wet-bulb thermometer. Assume that the initial temperature is T, wet-bulb temperature T_W, and mixing ratio w gm kg⁻¹. One kilogram of air releases

$$1000c_p(T - T_W) \text{ cal}$$

where c_p is the specific heat of the air. The heat required for evaporation is

$$L(w_w - w) \text{ cal}$$

where w_w is the saturation mixing ratio at T_W and L is the latent heat of vaporization. In the processes around the thermometer, the heat used for evaporation comes from the air. Therefore, these two quantities must be equal, that is,

$$1000c_p(T - T_W) = L(w_w - w) \tag{3}$$

If T and T_W are known, T_D the dew point can be determined. Let $T = 80°F = 26.7°C$, $T_W = 60°F = 15.6°C$, $w_w = 11.2$ gm kg⁻¹, $c_p = 0.24$ cal gm⁻¹ °C⁻¹, and $L = 586$ cal gm⁻¹. Substitution into Equation 3 gives $w = 6.7$ gm kg⁻¹. The dew point is then 8.1°C = 47°F. Computations usually give the dew point to the nearest degree only.

For temperatures below freezing, the wet bulb is covered with a thin coating of ice. Under these circumstances, the readings must be made very carefully because a small error in observing the temperature makes a big difference in the calculated dew point.

The cooling around a wet bulb is similar to that in other places. Perspiration and the water on a man at the end of a swim evaporate. In these situations, most of the heat required comes from the human body, cooling it. The cooling that comes from wetting a hot roof is an effect of evaporation of the water, and not of the water's original temperature. One of the reasons a field of vegetation on a summer day is cooler than a neighboring sandy area is that energy from the sun is being used to evaporate moisture coming from the plants and soil and, hence, does not heat the leaves or the air. These are only a few examples of ways in nature or under man's control where evaporation promotes cooling.

Two other methods of measuring humidity are in common use. The hair hygrograph uses the observed fact that organic fibers, such as human hair, increase in length with increasing relative humidity. This increase is recorded by a pen at the end of a pointer writing on a revolving drum, permitting a record to be made of the changes in the relative humidity (see Figure 2.10). The instrument is not highly accurate, and it must be checked against more accurately determined values.

In radiosondes and other instruments, the humidity is determined by measuring the electrical conductivity of a hygroscopic compound. The changes in the resistance can be relayed to a ground radio receiver, where the data are converted to relative humidities, and from them to the dew points or the mixing ratios of the air.

The dew point of the air changes when there is either evaporation or condensation. Normally, in fine weather evaporation is rapid during the warm part of the day, and the dew point rises slowly. Turbulent currents carry some of the additional moisture aloft so that the increase is not as much as might be expected. At night with dew or frost, the dew point falls. But these changes are slight in comparison with those of the relative humidity (see Figure 2.10). The rising temperature in the morning more than compensates for any increase in moisture, and the relative humidity falls, to increase once again when the temperature falls in the evening. In moist climates, the relative humidity usually reaches a maximum of more than 90 per cent during the latter part of the night.

4.4 Condensation of Water Vapor

If the temperature of the air, cooling, falls below the dew point, some of the moisture will change from the vapor state to a liquid or solid state. The dew or frost that falls is an example of this moisture which the air has lost. Some of the dew found on plants in the early morning is moisture that has come from the plants and has not evaporated and recondensed (see Exercise 12). When the cooling of the air occurs away from the earth, the excess moisture condenses as small water droplets or ice crystals. They are so small that they remain suspended in the air, forming clouds.

There is another common method by which air can become saturated. If cool air blows over a warm water surface, as for example during an autumn evening, evaporation from the water provides more moisture

FIGURE 4.4 Steam fog over Lake Erie. (Published by permission of Toronto Daily Star.)

than the air can hold. The excess again condenses in small water droplets, giving the *steam fogs* that can be seen under these circumstances (see Figure 4.4). The name comes from the fact that the process is similar to the one which produces the cloud of steam above a boiling kettle.

Careful study of cloud droplets has shown that they are not pure water. The moisture that condenses out of the air collects around small particles, called *condensation nuclei,* that are found floating in the atmosphere. They frequently are small crystals of salt, size 0.1 to 1 micron, that have come from the spray of the ocean. Other hygroscopic salts, such as ammonium sulfate, are also found to act as condensation nuclei. Another substance that may act as a condensation nucleus is sulfur dioxide, SO_2, which, oxidized by the sunlight, becomes sulfur trioxide. The water from the atmosphere combines with these particles to form sulfurous and sulfuric acid, respectively. After the air has reached saturation, the excess water collects on the largest of the nuclei present. It has been found that, with further cooling, condensation occurs on smaller nuclei instead of on the already-formed droplets. Thus the cloud consists of a large number of small water droplets. The sizes range from a few microns to 40 μ, with the most common size ranging from

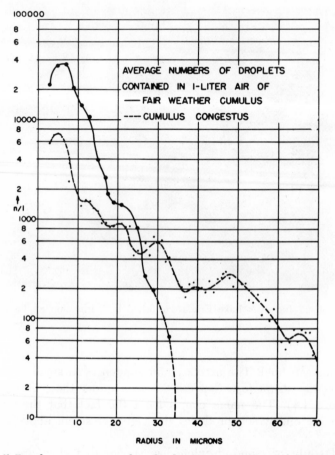

FIGURE 4.5 Droplet sizes in cumulus clouds. (From Weickmann, Helmut K., 1957: A nomogram for the calculation of collision frequencies. *Proc. First Conf. on the Phys. of Clouds and Precipitation Particles.* New York, Pergamon Press, 161–166.)

5 to 15 μ. Figure 4.5 gives the relative frequency of cloud droplet sizes in a heavy cumulus cloud. Cloud droplets of these sizes fall so slowly against the resistance of the air that they can easily be retained in the atmosphere. The largest of them fall slowly in a fog to produce *drizzle,* but the deposit from them is usually very small.

Even at temperatures as low as −40°C, water vapor will condense on condensation nuclei as water droplets. Ice crystals form around other small particles, *freezing* or *sublimation nuclei,* many of which come from the soils, rising as dust in the atmosphere. Freezing also occurs in some of the droplets when they collide with one of these nuclei.

When a few ice crystals are formed in a water-droplet cloud, they grow in both size and number, so that in ten minutes this cloud can change to one that contains only solid particles. In part, this process is accelerated because the vapor pressure over ice is less than that over water and, consequently, the situation is unstable when both are present. The ice crystals will absorb moisture from the atmosphere and will grow while evaporation occurs from the droplets because the air is no longer saturated with respect to liquid water.

4.5 Fog

Condensation of water vapor can occur near the ground, forming fog. The meteorological definition of *fog* specifies that water droplets reduce the visibility at eye level below 1 km ($\frac{5}{8}$ mi). The figure is arbitrary, because the effects of fog on an indivdual depend on the type of work he is doing. A coastal fisherman would find no difficulty in continuing work at visibilities which would stop an aircraft pilot from operating. The "official" definition is useful since it permits a comparison of fog frequency from two widely differing locations, eliminating any subjective judgment in the classification.

Commonly fog comes when the cooling at night reduces the temperature of the surface layer of air below the dew point. This fog is usually quite thin, being limited to the lowest 200 m, and sometimes, as in Figure 4.6, to the lowest 20 m. This type of fog is given the name of *radiation* or *ground fog*. It occurs with clear skies and light winds, permitting extreme nocturnal cooling. A layer of cloud aloft, reducing the outward flow of heat, inhibits its development. The spread between temperature and dew point is significant. Desert regions, even though the sky is clear, rarely experience ground fog because the temperature seldom falls sufficiently to produce saturation. Even on a calm night with moist air, fog does not always form because the excess moisture may be deposited on the ground in the form of dew. A light wind stirs up the air near the surface and, hence, permits the cooling to be distributed through 10 m or more. Higher winds spread the cooling through too thick a layer to permit the air temperature to reach the saturation point. When the temperature is much below freezing, the amount of moisture that condenses out (usually as ice crystals) is small, and the ice fog is thin. Radiation fog sometimes forms rapidly if a cloud deck moves away during the late afternoon, possibly after rain. The cloud deck will have kept the temperature low and the surface air close to saturation. With the clear skies, cooling by radiation reduces the temperature to the dew point, and fog forms.

FIGURE 4.6 Ground fog in a valley. (Published by permission of R. S. Scorer.)

Nocturnal cooling often gives rise to currents of cool air flowing down a hill to the valley floor (see Section 7.13 and Figure 7.22). It is in this cooler air that the fog is most likely to form, so that ground fog tends to be more common in valleys than on the adjacent slopes. Ground fog disappears rapidly in the early morning when the heat from the sun raises the temperature once again. Yet there are valleys in California and elsewhere where the ground fog tends to persist. Because the rays of the sun are reflected by the top of the cloud and hence do not heat the cooled layer, the fog persists during the day and is augmented with the cooling of the following night. The valley fog remains until a change in circulation sweeps the moist air away, replacing it with drier air.

The formation of *steam fog* has been described in Section 4.4. A common place for its formation is over the open waters of the polar regions in winter. Here the water temperature is at the melting point of sea ice (that is, −2°C), although the air temperature may be in the vicinity of −30°C. Evaporation from the open leads in the ice is rapid, and the acquired moisture saturates a large volume of the air that flows over it. This type of fog, because of its source, is given the name of *arctic sea smoke*. Steaming over water depends chiefly on the depression of the air temperature below the water temperature, although the humidity of the air and the salinity of the water are both significant. Under normal arctic conditions the steam fog from an open lead is widespread and persistent because the air usually is close to saturation even before the evaporation. Of course the evaporation uses heat from the water, cooling it. This may cause the water to freeze, and so cut off the source of moisture for the air. In some regions of the arctic the movements of ice and the flow of water under the ice keep replacing the cooled surface water, and the open leads can persist even with this loss of heat.

Saturation through evaporation may occur when warm rain falls through a cold layer of air near the surface. This sometimes occurs when two masses of air with different temperatures lie close together. It is therefore called a *frontal fog* (see Section 10.6).

Cooling of air below the dew point sometimes occurs when air is carried up a slope by the wind. The cooling by the reduction in pressure is considered in Sections 6.1 to 6.6, where the level of saturation is defined as the lifting condensation level. At this level a cloud forms, but an observer here will have his vision obstructed by the cloud and will call it a fog. The meteorologist classifies it as *upslope fog*. Upslope fog is common in coastal regions where the prevailing wind is on shore, such as coastal Oregon, Washington, and western Scotland.

The other major type of fog, *advection fog*, forms when a wind carries moist warm air across a cool area. The cooling of the air reduces the temperature below the dew point, and a fog results. Advection fog, because of its cause, is less influenced by the diurnal rhythm of temperature than ground fog. Cold currents in the ocean are a common cause of such cooling, and so advection fogs are frequently associated with them. For example, fogs are common along the coast of northern California and on the Grand Banks of Newfoundland, because warm air is carried over a cool water surface.

Although it is possible to identify various causes for fog and from these causes to classify them, frequently fog forms from a combination of causes. For instance, warm moist air may move inland from the Gulf of Mexico during the winter. The cooling over cold ground will reduce its temperature considerably. Further cooling will occur as the air rises over the slopes of the Appalachians. In this situation, its relative humidity can be so high that the diurnal rhythm of temperature carries it below the dew point during the night and fog results, to dissipate the following morning. The ultimate cause of the fog was diurnal cooling, but advective and orographic cooling both contributed to its formation. Figure 11.8 shows the weather map for 3 June 1962, 0100 EST, at which time Atlanta, Georgia, elevation 975 ft, had a dense fog from a combination of causes.

4.6 *Clouds and Cloud Forms*

When the condensation of water occurs above the surface of the earth, the result is a cloud. The usual cause of cooling is expansion in an upward movement of the air. These vertical currents may be slow, of the order of magnitude of centimeters per second, as in the area of an extratropical cyclone (described in Section 10.5); or they may be rapid, of the order of magnitude of tens of meters per second, as in a severe thunderstorm or tornado. Also, they may extend over a region hundreds of kilometers across, or may be limited to a small column, one to two kilometers across. The character of the vertical current is significant in determining the type of cloud that forms.

The classification of clouds is not simple. In 1803, a London pharmacist, Luke Howard, following the botanist Linneaus who classified plants, prepared a system of classification. This included families, genera, and species. Unlike plant types, a cloud may change its characteristics during a period of hours to develop into another type. A detailed classification will then have little value. A simplified version of Howard's classification is now in general use.

Clouds are classified by the altitude at which they form and, also, by the method of formation. There are three classes based on height, as follows.

1. Low clouds, base below 2 km.
2. Middle clouds, base between 2 and 6 km. A prefix "alto-" distinguishes these clouds.
3. High clouds, base above 6 km. They are identified by the prefix "cirro-" or the name "cirrus."

Two methods of formation divide clouds into stratiform clouds and cumuloform clouds. The former develop with slow and widespread vertical currents. Strong vertical currents over a small area produce cumulus-type clouds. These clouds can appear to be much deeper than their horizontal dimensions, but this is an error of perspective.

Out of the two types of classification, ten major cloud types are recognized. They are:

High clouds:

    1. Cirrus

    2. Cirrostratus

    3. Cirrocumulus

Middle clouds:

    4. Altocumulus

    5. Altostratus

Low clouds:

    6. Nimbostratus

    7. Stratocumulus

    8. Stratus

Clouds of vertical development:

    9. Cumulus

    10. Cumulonimbus

Consider first the family of stratus clouds. When air ascends slowly over a wide area, the cloud base is usually uniform in height and appearance. The cloud is then a *stratus* or layer *cloud.* Lacking contrasts, it fails to reveal its characteristics readily in photographs (Figure 4.7). In this group are the fog layers that have lifted off the ground, and

FIGURE 4.7 Stratus cloud or hill fog. (Published by permission of R. S. Scorer.)

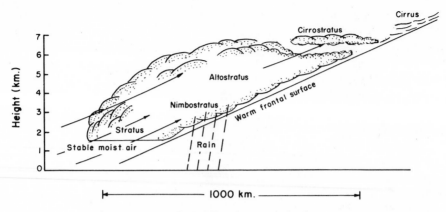

FIGURE 4.8 Family of stratus clouds.

the clouds formed when a layer of moist air is carried upward on a hillside. Occasionally, the wind breaks up the layer somewhat, producing fractostratus cloud.

When the rising currents are deep and the upper winds carry the rising air far ahead of the source, clouds may be spread over hundreds of kilometers. Near the source, these clouds are stratus (see Figure 4.8), but farther ahead the base has risen so that to the observer on the ground they are *altostratus* or, if rain is falling from the cloud, *nimbostratus*. Figure 4.9 shows that altostratus does not necessarily show a uniform base. At higher altitudes, the altostratus becomes thinner, permitting the sun to be seen through the clouds, and forming thin altostratus. The tops of the altostratus can be blown ahead of the main cloud deck. They are usually white, ice-crystal clouds which are so thin that they do not obscure the sun. Sometimes, these *cirrostratus* clouds are apparent only because they reduce the blue of the sky. At times a halo forms in the cirrostratus around the sun or moon. At other times, high clouds form in streaks or filaments in the sky, producing *cirrus* clouds (Figure 4.10).

The development of the family of stratus clouds, described above, may be interrupted through the effects of a *wind shear,* that is a change of wind with height, or radiation that causes vertical currents to develop within the stratus cloud. A stratus cloud may appear to have a smooth base even when an aircraft pilot will see many variations across its top. If the vertical currents that form these humps and valleys become sufficiently great, the varying thickness will become apparent to the observer underneath, who will identify it as a *stratocumulus cloud* (Figure 4.11) or a layer of cumulus cloud. The elements of the layer of stratocumulus may be joined, or may be separated with blue sky showing. Aloft, the altostratus changes its characteristics through the formation of *altocumulus*. A wind shear can cause the altocumulus to form itself into parallel bands of puffs of cloud (Figure 4.12). Altocumulus clouds are also seen downwind from a mountain ridge, forming in the tops of standing waves that form in the wake of the ridge (see Section 7.13). In a similar manner but less frequently, *cirrocumulus* clouds can form from a layer of cirrostratus.

The family of *cumulus* clouds provides the most variety and the greatest contrasts of light and shadow for the photographer. In these clouds, the vertical current producing the cloud has a diameter in the 1-to-40-km range. As described in Section 6.6, the height to which these currents rise is determined by the vertical temperature distribution. The base of a cloud occurs at that point where the air reaches the condensation level. Cumulus of fine weather, cumulus humilis (Figure 4.13), occurs if the cloud top is 1 or 2 km above the base, but a thicker cloud is known as heavy cumulus. In these circumstances, the cloud top is usually above the freezing level, but the condensed water is in liquid form and the top has a hard, cauliflower appearance. If one watches the top of a *heavy cumulus* cloud he will see turrets forming and dissipating within half an hour. They are then replaced by other turrets in other

FIGURE 4.9 Altostratus cloud. (Published by permission of R. S. Scorer.)

FIGURE 4.10 Cirrus cloud. (Published by permission of R. S. Scorer.)

FIGURE 4.11 Stratocumulus cloud. (Published by permission of R. S. Scorer.)

FIGURE 4.12 Altocumulus in bands. (Published by permission of R. S. Scorer.)

FIGURE 4.13 Cumulus of fine weather over London. (Published by permission of R. S. Scorer.)

parts of the cloud. If ice crystals begin to form within the cloud top, this changes its character to a filmy nature. Frequently this initiates a shower, and so justifies the name *cumulonimbus* (Figure 4.14). Although the top is frequently not seen, showery weather, or signs of extremely turbulent currents, gives evidence that a cumulonimbus cloud is above. These are the clouds of thunderstorms, hailstorms, and tornadoes. The tops are usually above 7 km, and frequently penetrate the tropopause.

There are a number of popular terms used to describe the amount of cloud in the sky. Official observation in North America reports the number of tenths of the sky covered with clouds. In some countries, the basic unit is one eighth of the total sky. Terms used in climatology are usually related to the cloud amounts as follows:

Less than 1/10 cloud cover: clear
1/10 to 5/10 cloud cover: scattered clouds or partly cloudy
6/10 to 9/10 cloud cover: cloudy
Over 9/10 cloud cover: overcast.

The term overcast is modified by the adjective thin when the cloud permits light from the sun or moon to pass through. Of course, it cannot be modified by the adjectives less or more.

4.7 Clouds as Seen by Other Means

Howard looked at clouds as he moved over the earth's surface and made his classification from what he saw. For more than 100 years, this was the only platform from which man could view clouds unless he looked down on a stratus cloud from a mountain top. Weather maps permitted the meteorologist to tie together the individual observations and to relate them to the pressure and wind distribution. Even then the platform continued to be the surface of the earth.

The advent of the airplane during the first quarter of the 20th century provided another platform for observing clouds. The aircraft pilot learned of the presence of clear layers within a cloud, of turbulence within cumulus clouds, of icing on his aircraft if he flew into some areas, and of cloud depth and cloud base, etc. The exchange of information between meteorologist and pilot added greatly to the former's knowledge of the weather. For instance, the frontal theory (Section 10.4), which was based originally on surface observations, was confirmed and modified with the aid of observations from aircraft. This theory unified the cloud structures around an extratropical low-pressure system.

The development of the radiosonde by 1940 permitted regular observations of the structure of the atmosphere and its moisture content. The data could be interpreted to give information on cloud locations, but it did not identify them clearly.

FIGURE 4.14 Cumulonimbus cloud at Pensacola Beach, Florida, 10 June 1961. (Published by permission of H. F. Materna. From *Weatherwise*, 1961, 14(4), Cover).

Radar, although invented with other uses in mind, has proved a tool for observing the three-dimensional structure of clouds. Actually a weather radar does not show the outline of the cloud but the volume where the drops are large enough to fall as rain. Figure 4.15*b* shows a radar picture of a thunderstorm taken at 1925 h MST 15 July 1968. Figure 4.15*a* shows a picture of the cumulus cloud taken at the same time. According to the sequence of radar pictures, the top of the echo rose from 25,000 ft at 1850 h to 33,000 ft at the time of the photograph.

FIGURE 4.15*a* Cumulus cloud at Penhold, Alberta, 1925 h MST 15 July 1968.

Thereafter, the top decreased to 24,000 ft at 2030 h. Further information on this storm is given in Sections 6.6 and 11.5. The use of radar has permitted meteorologists to study the cloud structure and its rapid changes in the kind of detail not possible by any other means.

Earth satellites have provided another platform from which man can view the clouds. Figure 4.16 shows a picture of the Atlantic Ocean taken at an altitude of 22,300 mi from the third Applied Technology Satellite on 10 November 1967. Portions of four continents may be seen. Clouds were thick over the Amazon Basin with a thin band stretching across the ocean to southwest Africa. Orographic clouds were visible along the west coast of South America where the air was being forced over the Andes range, and an extratropical storm lay off the east coast of Argentina with clouds along a cold front (see Section 10.7) which extended northwestward. Another cyclonic storm lay over Iceland, at the edge of the photograph, with a trailing band of cloud along another cold front in the western North Atlantic. Clear areas were also present. The features in this picture could be associated with the weather conditions existing at that time and could provide help in the analysis. Figure

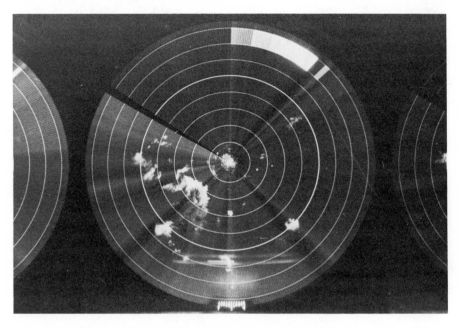

FIGURE 4.15*b* Radar picture of the cloud, taken at the same time. (Both pictures published by permission of Research Council of Alberta.)

4.17, with a much larger scale than Figure 4.16, shows three hurricanes in the North Atlantic on 17 September 1967.

Satellite pictures have confirmed many ideas about clouds that had previously been deduced. They have also brought out clearly some features that had not been well known. For instance, the pictures have shown bands of clouds spiralling inward toward the center of a hurricane (see Figure 4.17, the northern hurricane), and also shown parallel bands of stratocumulus clouds over ocean areas.

FIGURE 4.16 Clouds over and around the Atlantic Ocean, taken by III-ATS, 10 November 1967. (Courtesy of NASA. From *Bull. Amer. Meteor. Soc.*, 1968, 49(2), Cover.)

FIGURE 4.17 Composite of ESSA II photos 17 September 1967, of the western North Atlantic, showing positions of three tropical storms. (Courtesy, NASA. From *Weatherwise*, 1968, 21(1), 17.

Radiation thermometers (Section 2.2) on satellites reveal the temperature of the radiating surface of the earth. With clear skies, this is the surface of the ocean or the land; with cloudy skies, the tops of the clouds. Because the cloud top is usually much colder than the earth's surface, it is possible to estimate the pattern of clouds from the temperatures as reported by the satellite. These data are not restricted to daylight hours, and so they can provide information on the cloud structure at night.

4.8 *Precipitation*

Precipitation is the term used to cover all forms of water that fall to the ground in either solid or liquid form. It does not include dew nor the water droplets in fog that may be intercepted by trees, etc., although these deposits supply significant amounts of moisture in some places. These phenomena, and others, are all included in the term *hydrometeors.*

Liquid water falls as either *drizzle* or *rain.* Drizzle droplets, less than 0.5 mm in diameter, fall only from stable layers of stratus clouds close to the surface of the earth. Raindrops vary in size from the largest droplets of drizzle to 0.6 cm in diameter.

Water in solid form falls in a variety of shapes. At temperatures below freezing, water vapor sublimes on special types of nuclei to form *ice needles.* Sublimation causes them to increase in size, usually in a hexagonal pattern, giving snow. Large snowflakes form by the union of a number of single crystals, usually at temperatures close to freezing. At extremely low temperatures, the original ice needles may fall so slowly that they seem to be suspended in the air. Ice needles can also fall from low stratus clouds at temperatures of $-10°$ to $-5°$C.

Granular snow or *snow pellets* are balls of packed snow usually less than 1 mm in diameter. They are formed in the turbulent currents of unstable air. Snow grains, the solid equivalent of drizzle, fall from stratus clouds.

At times, rain falls from a warm layer aloft through a layer of air below freezing (see Figure 4.18 and Section 10.4). The raindrops become supercooled and turn to ice on collision with solid objects such as aircraft, trees, wires, or ground. This is *freezing rain.* The weight of the ice on trees, power lines, etc., can cause widespread damage as can be seen in Figure 4.19. If the rain freezes before reaching the earth, it becomes *frozen rain.* In eastern North America this is sometimes

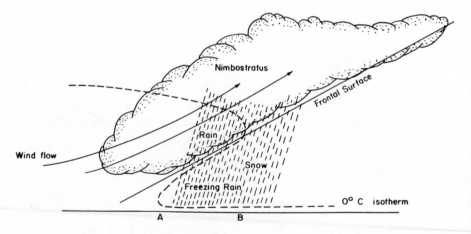

FIGURE 4.18 Development of freezing rain.

FIGURE 4.19 Trees covered by ice from an ice storm, Huntsville, Alabama, 2 March 1960. (Published by permission of Farley Vaughn. From *Weatherwise,* 1960, 13(5), 199.

called *sleet*. In the British Isles, where freezing rain is not prevalent, *sleet* is used to designate the mixture of rain and snow formed when snow falls through a layer of air above freezing and some melts.

Hail is formed only in massive cumulonimbus clouds. Vertical currents carry large amounts of moisture aloft. At the top of these clouds, pellets of soft hail are formed by the collision of the snowflakes found there. These, either through collisions with water droplets or by sinking below the freezing level for a short period, acquire a coating of liquid water which turns to ice. The growth of the stone can continue through two or three cycles, its support coming from the high vertical velocities of the air. Finally, it becomes too heavy to be supported, or it is thrown out ahead of the cloud to an area where the vertical currents are missing, and it falls to the earth. Most, but not all, hailstones show concentric patterns of growth. Figure 4.20 shows the size distribution of hailstones in New England, Colorado, and Alberta. Hailstones 10 cm (4 in.) in

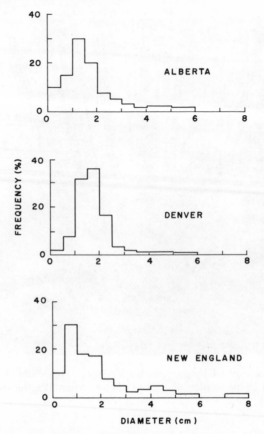

FIGURE 4.20 Size distribution of hailstones in New England, Colorado, and Alberta. (Published by permission of R. H. Douglas. From Douglas, R. H., 1963; Recent hail research, a review. *Amer. Meteor. Soc., Meteor. Monographs, No. 5,* 158.)

diameter have been found, and one with a circumference of 17 in. (13 cm diameter) fell in Texas in April 1944.

The measurement of rain does not present many complications, although care must be taken in choosing a site. The object is to determine the average amount that falls on a unit area. With snow, the difficulties are magnified, particularly because of the wind. The eddies around the mouth of a gage are such that it catches less than it should. The varying density of snow adds another complicating factor. This may be as low as 0.06 gm cm^{-3} with light fluffy snow, or as high as 0.3 gm cm^{-3} in arctic regions. On the average, 10 cm of snow has the moisture equivalent of 1 cm of rain.

Cloud droplets that form on condensation nuclei after the air has cooled to the dew point are too small to fall to the earth as rain. Their size increases very slowly as more water condenses with further cooling. Most of the additional water forms more drops of approximately the same size. Thus rain does not fall merely because the temperature of the air is reduced below its dew point.

A physical process by which cloud droplets could increase in size to form raindrops was first proposed by Wegener in 1911 and was developed further by Bergeron in 1933 and Findeisen in 1934. When both water droplets and ice crystals are present, the air which is saturated with respect to water is supersaturated with respect to the ice. This leads to a deposit of vapor on the ice crystals and, then, evaporation from the water to try to keep the air saturated. Thus the ice crystals grow rapidly and the liquid water tends to disappear, a result that has been observed within clouds. These larger particles fall more rapidly than the smaller ones and collect other particles that lie in their path, to make them even larger. Below the freezing level they melt, and continue to increase in size by collision as long as they are within the cloud.

In tropical and equatorial regions, rain falls at times from clouds whose tops remain below the freezing level. Here the Wegener-Bergeron-Findeisen process cannot operate, but a similar process seems to initiate the rain. The turbulent currents within the clouds at times bring into close proximity droplets of different temperatures or different sizes. The colder drops tend to grow at the expense of the warmer ones. Also, when the droplets differ sufficiently in size so that there is a significant difference in their rates of fall, collisions cause the large ones to grow larger. When the cloud is deep enough, the growth is sufficient to produce raindrops.

The study of the processes that produce rain has been stimulated by a hope that mankind would be able to increase the amount of water that falls from the clouds. If we can understand the physical processes that occur within clouds, it may be possible to modify them to obtain rain that would not otherwise fall.

4.9 Rain Making

The Wegener-Bergeron hypothesis on the formation of rain was confirmed by Langmuir and Schaefer in 1946. Into a supercooled water-droplet cloud in the laboratory they dropped some dry ice (carbon dioxide snow). Immediately some of the water droplets froze. In turn, they collected further moisture and fell to the bottom of the box as snow. Thus these scientists showed that a cloud could be dissipated and precipitation made to fall by man-made processes. Subsequently, they tested their experiment in a layer of stratus cloud out-of-doors. This cloud dissipated and precipitation in the form of snow fell from the base of the cloud. Later it was discovered that silver iodide crystals would act effectively as ice crystal nuclei and initiate the same set of reactions.

Since the initial experiments, many have been run. Some have tested the conditions under which the process worked and others have been conducted with the intention of using the process to increase rainfall. The latter tests have not been easy to evaluate. Because the conditions under which the seeding of clouds by silver iodide can be considered practical are the ones where rain might occur naturally, it is not sufficient to observe that rain fell from a seeded cloud and not from a nonseeded cloud. A comparison of precipitation amounts must be made with a recognition of the variation in rainfall in space and time. The current evidence suggests that rainfall caused by lifting moist air up a mountain-side (orographic rain) may be increased by about 10 per cent by cloud seeding. Experiments with cumulus clouds to produce convective rain have failed to give consistent evidence that the seeding has increased the rainfall.

Some exponents of cloud seeding have made claims that the seeding can be used to suppress hail. By injecting silver iodide into the cloud at low levels, they initiate a large number of ice crystals and, hence, remove the liquid water necessary for the growth of large hailstones. Clouds seeded by aircraft do show a rapid change from water-droplet clouds to ice-crystal clouds. Also there are reports that the solid deposits from these clouds are at times "soft hail," lacking the solid ice usually found in hailstones.

Recent experiments have suggested that the hail can be suppressed or the damage reduced if the layer just above the freezing level is saturated with the seeding agent. Scientists of the USSR report success in controlling hail in the Caucasus by exploding rockets with silver iodide near the —5°C isotherm within the cloud, thus supersaturating the core of the cloud. Testing is continuing in order that meteorologists may learn where and how best the seeding agent should be released within the cloud and the amount necessary to be effective. Some results have suggested that underseeding will increase the risk of hail and hail damage.

Theoretical studies have suggested that cloud seeding might be used to alter the structure of hurricane clouds sufficiently to reduce the damage caused by the wind. But the testing here is fraught with danger, and as yet no firm conclusions have been reached on the hypothesis.

4.10 *Visibility*

The distance at which objects can be distinguished is important to meteorologists. *Visibility* is defined as the average of the farthest distances at which objects can be identified, the average being taken for all directions. Nighttime visibility is measured by the distance at which lights of a given power can be seen, and these data are translated to give the visibility that would be present during daylight hours.

On a worldwide scale, the major cause for reduced visibility is condensed water in the form of fog. Depending on the thickness of the fog, the visibility may be reduced only slightly or may be so low that objects 100 m away are not seen.

Rain reduces the visibility only slightly, and the reduction with drizzle is associated more with the surrounding cloud than with the falling water. Snow and blowing snow both can reduce visibility. The meteorological definitions of light, moderate, and heavy snow are based on this reduction. High winds will pick up sand, dust, or snow from the earth's surface, carrying it along in the air and, thus, reducing the visibility. The effect with sand depends on the size of the particles, their looseness, and their dryness, as well as the strength of the wind. The same variables enter into the intensity of the blowing snow. With snow there is also a dependence on temperature. With temperatures near freezing, snow is too damp to drift much. In the polar regions, the snow structure demands a higher wind to lift the material from the earth than the flakes of more temperate latitudes.

Smoke is another cause for reduction in visibility. When the air is humid, some products of combustion act as condensation nuclei. The resulting fog combines with the smoke to form *smog* which is worse than either the smoke or fog that produce it. Smog is particularly serious around big cities like London, Los Angeles, and Pittsburgh. When the air is warmer aloft than it is at the ground (see Section 6.9), the smoke or smog remains trapped below the warm air and, consequently, around cities the visibility continues to decrease until a change of wind direction or speed removes the polluted air.

Very small solid particles in slight concentration, formed from dust, salt, smoke, etc., can at times reduce the visibility without revealing their identity. With high humidity, water will condense on them, causing a *haze* over the surface of the earth.

4.11 *Evaporation*

Most of the water enters the atmosphere by direct evaporation from the ocean, from other water surfaces, and from snow and ice fields. Transpiration from plants provides a smaller, but significant, amount of water to the air. The total passing into the atmosphere by direct evaporation and through transpiration is called *evapotranspiration*. During a period of rapid growth, the loss of water from a field of vegetation can exceed the loss from an equal area of a neighboring lake. The meteorologist is interested in the evapotranspiration because he must be aware of the moisture present in the atmosphere and its changes. Also, the evaporation uses much of the solar energy and so enters into the heat balance, as discussed in Section 1.10. Evaporation is more important to the hydrologist, who is concerned about the losses of moisture from streams and reservoirs, and to the agriculturalist, who must think of the needs and sources of water for his plants.

Determination of evaporation is not easy. Carefully run experiments in controlled conditions can give the rates of evaporation, but the instrument itself at times makes the reading somewhat artificial. The nearest approach to natural conditions is found with a *lysimeter*. A large volume of soil, disturbed as little as possible, is removed and placed in a large pan. The soil is then replaced carefully into its original position and is kept with the same vegetative cover as the surroundings. Scales below the pan keep a record of the changes in weight. Because other changes (rainfall, water from irrigation, overflow water, etc.) can be measured, the balance gives the evaporation from the soil and transpiration from the vegetation. Although lysimeters provide a measurement of evapotranspiration in a natural setting, they cannot be used with all varieties of vegetation, such as shrubs and trees.

Evaporation pans, large pans kept filled with water, have been installed to attempt to measure evaporation. Both installation and measurement of water loss are relatively easy compared to the ones with a lysimeter. On the other hand, absorption of heat by the pan and the water can raise the temperature above that of the natural surfaces, causing increased evaporation, and the wind at water level is slightly higher than at ground level. For these and other reasons, the measured evaporation from a pan is about 30 per cent greater than the evaporation from a lake in the vicinity, and neither one gives directly the evapotranspiration from an area with a vegetative cover.

Micrometeorologists have studied the theory of the flow of water vapor away from the earth's surface, and from their studies have developed equations giving the evaporation as a function of the vertical gradient of moisture and wind. Observations by lysimeters have permitted the theoretical development to be checked and the constants to be evaluated. Dalton developed one formula that can be used readily with standard meteorological data. It states that the evaporation E is given by the formula

$$E = K(e_s - e_a)u \tag{4}$$

where u is the wind speed, e_s the vapor pressure above the water, and e_a the vapor pressure in the air. K, a factor of proportionality, depends on the units used and also varies with the wind speed and the height of observation. A formula that gives reliable values for the evaporation over several mid-latitude lakes has the value of $K = 0.175$ mb^{-1} when u is the wind flow in cm at 2 m. Equation 4 gives the evaporation in cm per 3 hr when the wind flow is taken for the same 3-hr period.

Penman developed an equation that has wider applications. The theory is based on the net flux of energy to the ground, and requires data on the incoming short-wave and outgoing long-wave radiation. It has

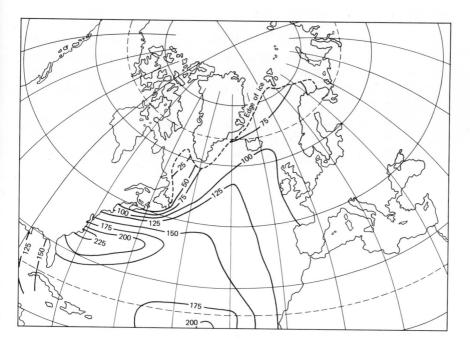

FIGURE 4.21 Mean annual evaporation (cm) over the Atlantic Ocean (after F. Albrecht, H. Baulig, M. I. Budyko, and others).

been used extensively in research problems but, because the necessary data are observed only at a relatively few locations, it is not so frequently used.

Figure 4.21 gives the mean evaporation for the year over the North Atlantic, as obtained from the work of a number of scientists. It represents approximately the distribution over other ocean areas as well. In some regions of the tropics, the rate exceeds 200 cm yr^{-1}, a value which is also found for the waters of the Gulf Stream off the southeastern United States. The rate changes rapidly as one moves from the Bermuda area to the cold waters off Newfoundland and Nova Scotia. The change occurs much more gradually in the central portion of the ocean. The variation over land areas is extreme, the amount being limited by the available water. In desert areas, evaporation equals or exceeds precipitation, the balance coming from inflow from more moist areas. The relationship between rainfall and evapotranspiration is used to delineate the desert and steppe areas of the earth from those areas where moisture is usually sufficient for plant growth.

PROBLEMS AND EXERCISES

1. By the method suggested in Section 4.3, paragraph 1, determine the dew point in several rooms of a dwelling. From your results, calculate the total weight of moisture contained in the building. (Below 2500 ft, the answer may be obtained with reasonable accuracy by using data from Appendix 2.)

2. Explain the presence of moisture, water or frost, on the windows of a building. Why should moisture form on some windows and not on others? In what manner does this deposit act as a control on the moisture content of the air?

3. A car windshield at times will become moist when moving rapidly from cold air to warm air. Explain. Under what conditions does this occur?

4. Determine the amount of moisture contained in a room 5 m × 4 m × 3 m at a temperature of 68°F and a dew point of 30°F

5. Saturated air at 0°F is carried into a house 15 m × 12 m × 10 m. and warmed to 70°F. How much moisture must be added if the relative humidity inside the house is to be kept at 35 per cent?

6. By using the method of Section 4.3, paragraph 3, determine the wet-bulb temperature (a) when the initial temperature is 15°C and the dew point −5°C, and (b) when the temperature is −5°C and the dew point −25°C.

7. A house 15 m × 10 m × 8 m is maintained at a constant temperature of 20°C and a relative humidity of 30 per cent. Outside air is saturated at −10°C. Under these conditions it is necessary to evaporate 1 kg of water per day to make up for the loss of moisture to the outside. What proportion of the air is exchanged per hour?

8. Take observations, twice daily for a week, of cloud types and amounts. On a day when the cloud cover is changing, increase the number of observations to six. Are clouds more prevalent at one time of day than another? Were you able to observe one type of cloud changing to another?

9. For a period of a week make observations of the occurrence or absence of dew. Notice, on days with dew, the areas where dew is found and the ones where it is not found. Explain.

10. On a morning after the temperature has fallen near the freezing point, observe the distribution of white frost on roofs and different ground surfaces. Is the variation a result of differences in temperature or in moisture?

11. When a vessel of water is placed on a stove to bring the water to a boil, it will heat more rapidly when it is covered. Explain.

12. Air near the surface of the earth is saturated at a temperature of 20°C. If all the moisture in the lowest 100 m is deposited as dew during the night, how much would be found on 1 cm²? (Assume a surface pressure of 1000 mb.) How realistic are the assumptions? What conclusions can you reach about the source of dew?

13. One cubic meter of air near the earth's surface has $p = 1000$ mb, $T = 10°C$, $T_D = 10°C$. Determine the mixing ratio. The air is now lifted to 500 mb, with heat added so that the temperature is kept constant. Determine the volume, mixing ratio, saturation mixing ratio, and relative humidity. What do your results indicate about the change of dew point with a change in pressure? (By using a tephigram or emagram, described in Section 6.4, one can determine the new dew point.)

CHAPTER 5

RADIATION

The subject of the energy received from the sun was introduced in Sections 1.2 and 1.5. It was shown that this quantity varied with latitude, and also that the heat lost by the earth varied. These variations result in an imbalance of heat which is corrected by the flow of air and water over the surface of the globe. It is now necessary to return to the subject of radiation and examine its characteristics. Also, we must examine more fully the results that arise from this method of energy or heat transfer.

5.1 *Methods of Heat Transfer*

Students of physics recognize three methods by which heat is carried from one place to another: conduction, convection, and radiation. All three are significant in weather processes but, of the three, radiation is fundamental.

In conduction, the heat energy is transferred from molecule to molecule within the substance. Thus the extra heat energy at one end of a rod moves to the other end and, as a result, the rod acquires a more nearly uniform temperature. Air is a poor conductor of heat and, hence, transfer of heat by conduction in the atmosphere is slow. Some transfer is made across the earth-air interface, but this affects only the air within a few centimeters of a plane surface or the air that surrounds the vegetation and other protuberances. Heat moves from the surface of the earth downward into the soil (and returns again later) by conduction.

Heat is transferred by convection when currents of fluid such as air or water carry heated masses from one location to another where they give up their store of heat. In meteorological processes, horizontal movements are sometimes designated by the term *advection,* reserving *convection* to describe transfer by vertical movements. The meteorologist uses these terms both when warm air is moving to cold areas and when cold air is moving to warm areas. Convection and advection are common on the earth's surface, bringing about the distribution of heat described in Chapter 1.

Heat transfer by radiation is in the nature of waves of energy passing through a medium without changing the medium. Radiation covers the transfer of energy by electromagnetic waves over a broad spectrum, as indicated by Figure 5.1. At one end are the cosmic rays and X rays. At the other end of the spectrum are the waves that carry our radio programs. Most important from the meteorologist's point of view are the waves in the wavelengths from 0.2 μ to 50 μ. The wavelengths from 0.4 to 0.8 μ have the property of exciting nerves of the human eye and, hence, are called the visible light rays. Of these, violet has the shortest wave length and red the longest.

Transfer of heat by radiation is familiar to all who have stood in front of a fire. Radiant heat passes in a straight line from the fire to the parts of our bodies facing the flame, while the other side of our bodies, receiving no heat from the fire, can be quite cold. The paths of radiant heat are known to us by our knowledge of light rays. They are generally straight, but can bend when passing into another medium such as water. Or they can be reflected by some surfaces, or can be partly absorbed and partly reflected as, for example, when sunlight hits a colored object. Long-wave radiation (wavelength greater than 2 μ)

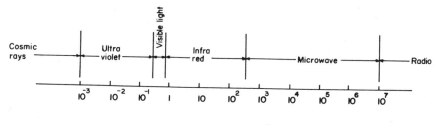

Wave length (microns)

FIGURE 5.1 The electromagnetic spectrum.

is less familiar to us, yet it plays an important part in the balance of heat on the earth.

5.2 Laws Governing Radiation

Every body in the universe radiates energy in the form of electromagnetic waves. The total amount of energy E emitted varies with the fourth power of the absolute temperature. The amount varies from body to body, but the maximum amount per unit area is given by the Stefan-Boltzmann formula

$$E = \sigma T^4 \tag{1}$$

where $\sigma = 8.13 \times 10^{-11}$ cal cm^{-2} deg^{-4} min^{-1}, the *Stefan-Boltzmann constant*. A body that radiates the amount given by the formula is called a *black body*.

Two other laws deal with radiation and the energy given off by a body radiating in all wavelengths. Planck's law deals with the variation in energy emitted over the different wavelengths. The energy distributions for 200°, 250°, and 300°K, derived from Planck's equation, are given by the curves of Figure 5.2. The radiant energy is represented by the area under the curves. The amount for 300°K is over five times the amount for 200°K.

The second law, Wien's Displacement law, says that the wavelength, in microns, for which the energy is a maximum (point M in Figure

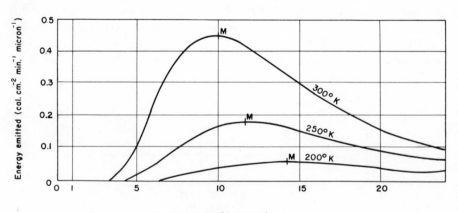

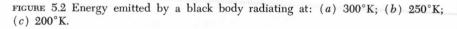

FIGURE 5.2 Energy emitted by a black body radiating at: (*a*) 300°K; (*b*) 250°K; (*c*) 200°K.

5.2) is given by $\lambda_m = 2900/T$. Notice that the wavelength for maximum energy decreases with increasing temperature. Thus a body such as the earth, which has a temperature of 68°F or 293°K, transmits the greatest amount of energy in the wavelengths near 10 μ, that is, in the infrared wavelengths. In radiation from the sun, where the temperature is near 5600°K, the maximum energy comes in the wavelength of 0.5 μ, that is, in the green part of the visible light band. Almost one half of the energy from the sun comes in the wavelengths of visible light. This is in marked contrast with the distributions shown in Figure 5.2. For this reason, solar radiation is frequently called *short-wave radiation,* although in reality radiation from the sun comes in all wavelengths. Most radiation from the earth and from bodies of about the same temperature has a wavelength near 10 μ and is called *long-wave radiation.*

Although the three laws are stated separately, they are closely interrelated. Planck developed a general equation that united the other two laws and other information then known about radiation. This equation gives the energy per square centimeter per minute per micron as

$$E_\lambda = c_1/\lambda^5(e^{c_2/\lambda T} - 1)$$

Here c_1 and c_2 are constants, and e is the base of natural logarithms 2.718. From this equation both the Stefan-Boltzmann relationship and Wien's law may be derived.

The opposite to emission of radiation is absorption. When radiant energy falls on a body, it may be transmitted, reflected, or absorbed. There is a selective process in the action of the body. Glass permits the rays in the visible light range to pass through with very little change, and we speak of glass as being transparent. Yet glass is selective and does not permit rays in the longer wavelengths to pass through. This selective absorption permits glass to be used in greenhouses. Solar radiation passes through the glass in the roof to be absorbed by the ground and plants. The long-wave radiation from them is absorbed by the glass, passes to the outside of the glass slowly by conduction, and then by conduction and radiation to the air. This trapping of the solar energy keeps the temperature of the greenhouse higher than the temperature outside. Similarly, flesh permits the very short X rays to pass through, although it blocks most radiation, and so we get X-ray pictures of the body.

Colored objects absorb rays in the visible light range selectively. A blue object absorbs less of the blue rays than of the other rays. The reflected light has then a predominance of blue rays, and the object appears blue.

5.3 *Measurement of Radiation*

The instrument most frequently used to measure radiation makes use of the laws of absorption and emission. In the Eppley pyranometer (see Figure 5.3), there are white and black sections. The absorption of solar radiation by the black sections is greater than that by the white sections, causing a temperature difference. This difference is measured by electrical means, and from it the amount of energy being received is determined. The same principle may be used to measure the flow of long-wave radiation from the earth but, in this case, one must examine the difference in the absorption of the long-wave radiation by the two rings.

5.4 *Solar Radiation*

The sun, a body with a surface temperature of about 5600°K, radiates most strongly in the visible light range. The amount of energy given off is very great. At the orbit of the earth, the energy is still 2.0 cal min^{-1} on every square centimeter perpendicular to the sun's rays. This amount of energy is the *solar constant*. One calorie per square centimeter is called a *langley*.

The gases of the atmosphere are almost transparent to these wavelengths and, therefore, much of the energy from the sun can penetrate to the earth's surface. One gas in the atmosphere is opaque to the wavelengths in the ultraviolet. When waves of length less than 0.3 μ meet

FIGURE 5.3 Pyranometer for measuring the radiation from the sun. (Courtesy, Eppley Laboratory Inc.)

oxygen molecules, these are changed to form ozone (Section 1.4). This occurs in the high atmosphere where, consequently, there is a continual manufacture of ozone during the sunlit hours. As a second result, the sunlight that reaches the surface of the earth is robbed of the radiant energy of wavelength below 0.3 μ. The absorption of this energy and the later release of it, as the ozone reverts to oxygen, is the cause of the warm layer in the atmosphere at the top of the stratosphere (see Section 2.7). The gases in the troposphere, particularly water vapor, absorb some of the short-wave radiation, causing a slight increase in temperature.

Some of the solar radiant energy is diverted from a straight path, that is, scattered by the atmosphere. When the air is pure, this scattering is more common with the shorter wavelengths. The scattered light reaches our eyes from all directions, and so we see the sky as blue. This effect is missing at high altitudes, and travelers in the upper stratosphere report the sky as black. Because the passage through the atmosphere robs the sunlight of some of its blue components, light at the time of sunset has an orange tinge. Light reflected by the base of clouds when the sun is below the horizon has moved even further toward the red end of the spectrum. Large particles in the atmosphere, dust, small water droplets, etc., scatter radiation in all wavelengths almost equally, and the sky then appears white.

Clouds affect solar radiation. Much of the radiation is reflected, not in the uniform manner of a mirror but diffusely. Notice how white the sunlit side of a cloud is in comparison with the dark side. Some of this radiant energy that hits the cloud passes back to outer space without having any effect on the temperature of the earth. A small amount of short-wave radiation is absorbed by the clouds, heating them.

What happens to the short-wave radiation received by the earth is summarized in Table 5.1. Of the radiation intercepted by the earth, on the average a little more than one half (51.5%) reaches the surface either directly or indirectly by scattering and reflection. A small amount (4%) is reflected by the earth's surface and escapes to outer space, again having no influence on the temperature of the earth. A fresh snow surface is the best reflector, reflecting about 80 per cent of the incident short-wave radiation. A smooth water surface is also an excellent reflector, particularly with a low sun. The total reflected by the earth and its atmosphere is about 35 per cent of the incident radiation. This fraction is called the earth's *albedo*.

Slightly less than one half the incident radiation is absorbed by the earth's surface. Some of the energy is stored as sensible heat, warming the earth and air, some is used to evaporate water, and some is used

TABLE 5.1 Annual World Average Transfer of Heat by Radiation and Other Means (in Per Cent of Incoming Solar Radiation)

Short-wave Radiation from Sun			
Incoming at the outside of the atmosphere			100
Reflected to space			
After atmospheric scattering		7	
From clouds		24	
From earth's surface		4	35
Absorbed by stratosphere		3	
Absorbed by tropospheric air		13	
Absorbed by clouds		$1\frac{1}{2}$	
Absorbed at earth's surface			
By direct beam	$22\frac{1}{2}$		
After scattering	$10\frac{1}{2}$		
After reflection by clouds	$14\frac{1}{2}$		
		$47\frac{1}{2}$	65
Heat Transfers Between Surface and Troposphere			
From earth to troposphere			
By conduction	11		
By latent heat of evaporation	$18\frac{1}{2}$		
By long-wave radiation	109		
		$138\frac{1}{2}$	
From troposphere to earth			
By long-wave radiation		$96\frac{1}{2}$	42
Long-wave Radiation to Space			
From stratosphere		3	
From troposphere		$56\frac{1}{2}$	
From earth's surface		$5\frac{1}{2}$	65

in the photochemical processes of plants and is stored in the resulting plant fibres.

5.5 Long-wave Radiation

As stated in Chapter 1, the earth must lose the absorbed solar radiation if a constant temperature is maintained. This is done by radiation from the earth's surface and from the atmosphere. Most of this energy is contained in waves with lengths in the range from 5 μ to 14 μ, that is, long-wave radiation.

The transfer of energy in these wavelengths is complicated because both carbon dioxide and water vapor tend to absorb radiation in certain bands. In general terms, these two gases absorb all radiation between 5 and 7 μ and above 14 μ, and some of the radiation in adjacent bands.

Between 8 and 10 μ, there is very little absorption in a clear atmosphere. Clouds tend to absorb all radiant energy in all wavelengths.

With clear skies, some heat is lost from the earth's surface directly to outer space. In the bands where absorption occurs, the energy is absorbed in the layer close to the earth's surface, warming it. This layer again emits this energy—some downward to the earth's surface and some upward to be absorbed in the next higher layer. Gradually heat is transferred upward until it reaches the top of the troposphere. Because the stratosphere has very little water vapor, the radiation upward from the top of the troposphere passes with only slight loss to outer space.

The short-wave radiation reflected by the earth is observable from satellites outside the atmosphere. The surface appears bright, since the earth reflects about one third of the short-wave radiation received from the sun. The brightest areas are the polar ice caps and the cloudy regions of the earth. There are also differences in the long-wave radiation. In those wave lengths in which the atmosphere is transparent, satellite instruments are able to detect radiation from the surface of the earth or from the tops of the clouds. An analysis of this radiation gives the temperature of the radiating surface. The instrument is similar to the radiation thermometer that produced the results shown in Figure 2.7. In the bands for which the troposphere is opaque, the molecules radiating energy to outer space are the water vapor molecules near the tropopause, radiating at the tropopause temperature.

A cloud radiates energy in all directions. The radiation from the base of the cloud is downward so that, with an overcast sky, energy is bounced back and forth between cloud and ground. With low clouds, the net exchange is small. Meanwhile, the top of the cloud is radiating more energy than it receives by downward radiation from the air above it. Thus the net effect of radiative exchange is to increase the lapse rate within the cloud, a result which is shown in Chapter 6 to be significant.

Even in the free atmosphere, a similar situation arises with radiation from the earth tending to increase the temperature of the lowest layers while the air at the tropopause gets colder. The net result, in general, tends to increase the lapse rate and, consequently, to decrease the stability. A further discussion of long-wave radiation and its effects in modifying the temperature of the free atmosphere is given in Section 8.3.

5.6 *Effects of Radiation on Temperature*

As described in Chapter 1, the total heat energy transfer by radiation provides the balance by which the temperature of the planet is kept constant. When one considers a specific locality or time, other factors become significant.

Consider Figure 5.4, which gives in curve (*a*) the mean temperature through the year at Washington, D.C. and in curve (*b*) the insolation at the top of the atmosphere at 40°N. At this latitude, the total heat gained from the sun is almost equal to the total heat lost to outer space.

Curve (*b*) is symmetric with respect to the solstices, being determined by the altitude of the sun. On June 22, the sun is at its highest point, and the heat received is a maximum, but the temperature is not. On June 22, the outgoing radiation is less than the incoming radiation; in other words, there is a net gain of heat warming the ground and the air, and the temperature continues to rise after this date. With the increasing temperature causing an increase in the outgoing radiation, and with the decreasing insolation, the two flows approach equality. During the third week of July, the flows balance and the normal date of maximum is reached. For the same reason, the date of minimum temperature is observed to occur not at the time of the least insolation from the sun, December 22, but four weeks later, near the end of January. The curves for Minneapolis, Minnesota and Fairbanks, Alaska (Figure 5.5) show much the same variation.

At times, the net gain or loss in heat is governed by factors other than the heat from the sun. In India, June sees the onset of the monsoon with its clouds and rain. Increased reflection from the clouds decreases the effective heat from the sun so that the maximum temperature comes

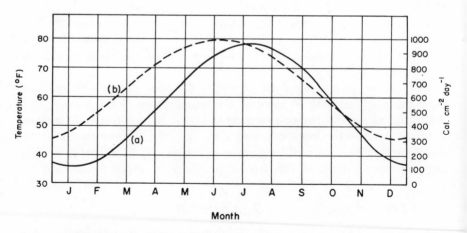

FIGURE 5.4 Washington, D.C.: (*a*) mean annual march of temperature; (*b*) mean solar radiation at the top of the atmosphere at 40°N.

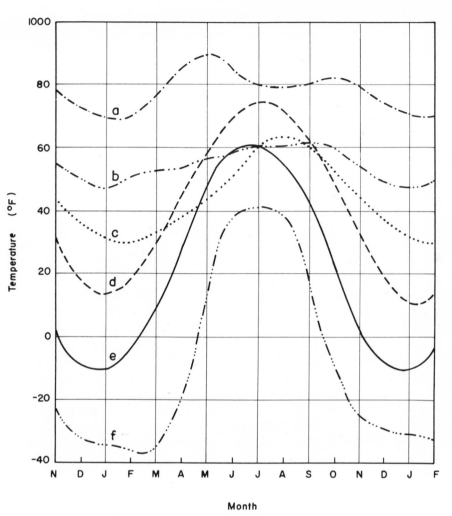

FIGURE 5.5 Mean annual curves of temperature: (a) Bombay, India, with a monsoon climate; (b) San Francisco, California, a tropical west-coast location; (c) Sable Island, Nova Scotia, a mid-latitude east-coast island; (d) Minneapolis, Minnesota, in a mid-latitude continental location; (e) Fairbanks, Alaska, with a polar continental climate; (f) Eureka, Northwest Territories, a station north of the Arctic Circle (80°N).

in May, before the time of maximum insolation. Also, in October, after the retreat of the monsoon, the effective insolation increases to halt the rate of fall of temperature and, at times, to cause October to be warmer than September. The annual variation for Bombay (Figure 5.5) shows this effect.

In marine areas, heat from the sun is used to warm a thick layer of water. Because of its high heat capacity, the temperature and, therefore, the outgoing radiation rises slowly. The points of balance occur, then, some weeks after the time of highest and lowest insolation. Island stations, such as Sable Island, Nova Scotia (Figure 5.5), find their coldest time of year in February and their warmest time in August.

The polar caps present another deviation from the normal annual trend of temperature. In the latitudes where the sun is below the horizon for a month or longer, the heat from the sun is cut off entirely for that period, and the amount of heat received during the first week after the sun first appears is small. Advection of heat from lower latitudes checks the rate of fall, but the temperature continues to drop during the polar night and does not begin to rise until the second or third week after the sun appears. The curve for Eureka, Northwest Territories, Canada (80°N 86°W) shows the effect of the polar night. At Eureka, the sun first appears above the horizon about February 23. The winter curve, obtained by a careful analysis of daily temperature data, shows a long slow decline in temperature from early December until early March. Such a winter has been given the name of *kernlose* or coreless.

The summer curve for San Francisco gives another example, similar to that at Bombay, of how a change in weather pattern can affect the temperature. In the summer months, a sea breeze from the Pacific keeps the city cool. This breeze decreases in strength in late August and, therefore, the solar radiation, even though down from the maximum, can become more effective, and the temperature rises to its highest in September.

The relationship between the annual temperature curve and the annual variation in solar radiation has points of similarity with the relationship between the diurnal variations of temperature and radiation. But there are minor points of difference.

On clear days with little change in weather, the warmest time of day is about two hours after the highest sun. At this time, the flows of energy to and from the surface of the earth balance. Some of the energy is passing into the soil, heating subsurface layers, and some is passing by conduction to the air. The major losses of energy are in long-wave radiation and in evaporating water, which increase with rising temperature. The air temperature at the level of the thermometer screen lags slightly behind that at the ground surface, but the time difference is small. The temperature curve at night is similar to the winter curve for the polar latitudes. The temperature drops rapidly from the time of maximum until about sundown. The decline becomes slower as the ground cools and because, frequently, the cooling results in condensation of water as dew to release heat. The time of balance comes when, after the sun has risen, the additional heat from the sun is sufficient to balance the radiational heat loss. This is about one-half hour after sunrise. With dry air, both the absorption of long-wave radiation and the back radiation from the sky to the earth are much less than with moist air. In the region covered by dry air, the heat loss during the night therefore reduces the temperature more than in regions where the air is moist, even though the sky is clear at both locations.

Clouds, too, affect the daily temperature curve. The cloud forms a blanket so that much of the solar radiation fails to reach the earth. At night, back radiation from the cloud reduces the net loss. This is observed in the temperature trace for 6 May in Figure 2.10. If clouds move into a locality about noon, then the daily temperature curve shows a maximum before the normal time. This corresponds to the annual curve for Bombay (Figure 5.5). In the same way, a change in the advection of hot or cold air during a specific day will cause the curve to deviate from the normal. The temperature curve for 9 May 1968, for Ellerslie, Alberta, given in Figure 2.10, shows an example of this. The sea breeze provides another good example; the resulting daily temperature curve has features similar to the ones of the annual curve for Bombay or San Francisco (Figure 5.5). The mean daily curve will show the same shape for places where the sea breeze blows regularly.

Although the average times of maximum and minimum do not alter greatly, the range between the two values varies for a number of reasons, including the ones already discussed. Increasing latitude, cutting down the insolation, and increasing humidity reduce the amplitude of the temperature curve. The type of soil and the soil moisture both cause differences. If the soil is a poor conductor of heat, radiant heat warms only a thin layer, but warms it much more rapidly than a soil which conducts heat downward. Also, if energy is being used to evaporate water either directly from the soil or through the leaves of plants, it does not become apparent as sensible heat. The daily range is greater over desert sand than over soil, and greater over bare soil than over grass or over swamps. When the daily range passes through the freezing point, it is reduced because the outgoing energy is partly derived from

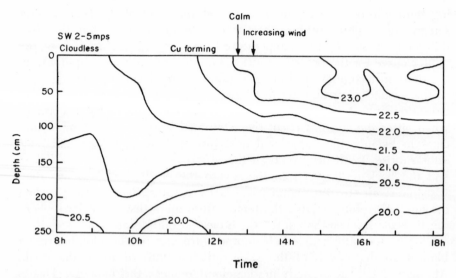

FIGURE 5.6 Course of temperature in a shallow lake, 17 July 1954 (after J. Herzog). (From Geiger, R., 1965: *The Climate near the Ground*. Cambridge, Harvard University Press, 4th edition, p. 192. Published by permission of Harvard University Press.)

the latent heat of the freezing of water, and incoming heat is used to remelt the ice.

Water has a relatively large heat capacity, thus reducing temperature changes. Evaporation uses up about one quarter of the energy absorbed by the earth, without becoming sensible heat. Over water areas the proportion used is much greater. In areas where dry air flows over warm water, the ratio may reach 80 to 90 per cent. An example of such an area is the Mediterranean Sea. During the period of cooling, the surface water loses its heat most rapidly. Then, cooler and denser than the water below, it sinks and is replaced by warmer water. Thus the effect of diurnal cooling is distributed over a layer of water several feet thick. Waves will carry the cooled water to even greater depths. The variation in temperature in a pond (Figure 5.6) shows how heat is transferred downward in a lake.

A similar effect is found in the atmosphere. When the wind is blowing at night, the air that has been cooled near the earth is carried aloft and replaced by warmer air. With little or no wind, the decrease of temperature at the earth's surface will be much greater because the air that is cooled is restricted to a relatively thin layer. Consequently, ground frosts tend to occur on clear still nights.

PROBLEMS AND EXERCISES

1. The temperature of the wires of a toaster increased from 800°C to 900°C. By what ratio is the emitted energy increased?

2. An iron ball is gradually warmed from 0°C to 1500°C. What changes have occurred in the emitted radiation to explain the terms "red hot" and "white hot"?

3. Sometimes a skier finds that he can ski without being uncomfortably cold with little clothing on the upper part of his body, even though the temperature may be 20°F or below. Consider the heat transfer processes that lead to this result.

4. On the basis of the processes discussed in this chapter, explain why an unventilated thermometer in the sun does not register the correct air temperature.

5. Assume that the earth is radiating as a black body at a temperature of 0°C. Because of a climatic change, the radiating temperature increased by 5°C. By what per cent would the heat lost by radiation increase?

CHAPTER 6

STABILITY

6.1 Heat, Temperature, and Volume Changes

Section 2.8 considered the effect of adding heat to a body. For a solid or a liquid where no change in state is involved, the equation

$$\Delta H = mc\, \Delta T$$

gives the relationship between the heat added ΔH and the change in temperature ΔT where m is the mass and c the specific heat.

Gases present an added complication because the volume may change. When this happens, some of the heat energy is used in doing work by expansion against the external pressure, or energy is acquired by compression. When the volume is kept constant, the same relationship

$$\Delta H = mc_v\, \Delta T \tag{1}$$

holds true, but now c_v must be specified as the specific heat at constant volume. When a change in volume ΔV occurs, consider the amount of change above one cm^2. A force equal to p acts through a distance Δd and, hence, does work that is equal to $p\, \Delta d$. When the work done over all unit areas is totaled, one discovers that the total work done is $p\, \Delta V$, or $m\, p\, \Delta v$ where Δv is the change of volume per unit mass. Instead of Equation 1, the conversion of energy is given by

$$\Delta H = m(c_v\, \Delta T + p\, \Delta v) \tag{2}$$

For unit mass

$$\Delta h = c_v \, \Delta T + p \, \Delta v \tag{3}$$

In using these equations, care must be taken with units. When p is measured in dyne cm^{-2}, $p \, \Delta v$ is in dyne-cm, or ergs. Heat is usually measured in calories. The conversion factor

$$1 \text{ cal} = 4.187 \times 10^7 \text{ ergs}$$

must be employed in order that the equation may be used.

Equation 3 may be transformed by means of the gas equation (Equation 4, Section 3.3)

$$pv = RT \tag{4}$$

Consider the situation when p, v, and T are altered to $p + \Delta p$, $v + \Delta v$, and $T + \Delta T$, respectively. Then

$$pv + p \, \Delta v + v \, \Delta p + \Delta p \, \Delta v = RT + R \, \Delta T \tag{5}$$

Subtracting Equation 4 from Equation 5

$$p \, \Delta v + v \, \Delta p + \Delta p \, \Delta v = R \, \Delta T$$

Now $\Delta p \, \Delta v$ is usually small compared with the other terms. Therefore

$$p \, \Delta v = R \, \Delta T - v \, \Delta p \tag{6}$$

Substitute from Equation 6 into Equation 3

$$\begin{aligned}
\Delta h &= c_v \, \Delta T + R \, \Delta T - v \, \Delta p \\
&= (c_v + R) \, \Delta T - v \, \Delta p \\
&= (c_v + R) \, \Delta T - \Delta p / \rho
\end{aligned}$$

When pressure is constant, that is, when $\Delta p = 0$

$$\Delta h = (c_v + R) \, \Delta T$$

and $c_v + R = c_p$, the specific heat with constant pressure. Therefore, in general,

$$\Delta h = c_p \, \Delta T - \Delta p / \rho \tag{7}$$

Although the combination on the right-hand side gives the total heat added, the second term has no simple physical meaning.

EXAMPLE: Consider the energy changes when an air-filled balloon, 1 m radius, temperature 290°K, pressure 1000 mb, is lifted to 850 mb, temperature 283°K. By using Equation 4, Section 3.3,

$$\rho_1 = 1.20 \times 10^{-3} \text{ gm cm}^{-3}$$
$$\rho_2 = 1.05 \times 10^{-3} \text{ gm cm}^{-3}$$
$$\Delta T = -7°C, \; \Delta p = -150 \text{ mb}$$

By using a mean density of 1.13×10^{-3} gm cm^{-3} and c_p from Table 2.2

$$\Delta h = 0.240 \text{ cal gm}^{-1} \text{ deg}^{-1} \times -7 \text{ deg} \times 4.187$$
$$\times 10^7 \text{ erg cal}^{-1} + \frac{150 \times 10^3 \text{ dynes cm}^{-2}}{1.13 \times 10^{-3} \text{ gm cm}^{-3}}$$

$$= 6 \times 10^7 \text{ erg gm}^{-1}$$

$$m = 1.20 \times 10^{-3} \text{ gm cm}^{-3} \times \tfrac{4}{3}\pi \times 10^6 \text{ cm}^3$$

$$= 5.0 \times 10^3 \text{ gm}$$

Total energy injected into the balloon is 3×10^{11} ergs.

This result is interesting. Although the temperature has dropped, presumably through a loss of heat, the work done to expand the balloon has taken up the heat lost by the gas and additional energy from an outside source. It becomes apparent that, when energy changes in the atmosphere are being considered, the changes in pressure may introduce quantities of the same order of magnitude as the ones involving changes in temperature. In the example quoted, the computations indicated a net gain in energy. This has probably come from the surrounding air. In the free atmosphere, a gain may be derived by a transformation of kinetic energy. This happens when a wind blowing against a slope of a hill starts the air parcel moving upward. Another source of energy for a parcel is the buoyancy force if the rising parcel is warmer than the surroundings. In the readjustment, air from aloft subsides to fill the space left vacant by the balloon, and the potential energy it loses in subsiding is taken up by the rising balloon.

When no change of energy in the parcel of air occurs, Δh in Equation 7 is 0 and

$$c_p \Delta T = \frac{1}{\rho} \Delta p$$

For the free atmosphere, the hydrostatic equation gives

$$\Delta p = -g\rho \Delta z$$

Therefore

$$c_p \Delta T = -g \Delta z$$

or

$$\frac{\Delta T}{\Delta z} = -\frac{g}{c_p} \tag{8}$$

This gives the rate at which the temperature drops with height when no heat is added from or lost to the outside. In numerical terms

$$\frac{\Delta T}{\Delta z} = \frac{-981 \text{ cm sec}^{-2}}{0.24 \text{ cal gm}^{-1} \text{ deg}^{-1} \times 4.19 \times 10^7 \text{ gm cm}^2 \text{ sec}^{-2} \text{ cal}^{-1}}$$

$$= -9.8 \times 10^{-5} \text{ deg cm}^{-1}$$
$$= -9.8 \text{ deg km}^{-1}$$

In round figures, this is considered 10 deg km^{-1} or 1 deg Celsius per 100 meters.

When a process such as the one described is carried out without gain or loss of energy to the parcel, it is called *adiabatic* (no change in heat). Many convective processes in the atmosphere occur so rapidly that little or no heat is gained or lost, and the processes may be considered adiabatic. In such a process, a rising parcel of dry air cools as it rises at the rate of 1 deg C per 100 meters or 5.4 deg F per 1000 feet. This value is called the *dry adiabatic lapse rate,* and is denoted by a capital Greek gamma Γ. Subsiding air warms at the same rate.

6.2 *Stability of Dry Air*

The vertical movement of air, discussed in the preceding section, introduces the question of stability in the free atmosphere. Physicists recognize three states of stability: stable, neutral, and unstable equilibrium. A cone on a table may be used to illustrate all three states (Figure 6.1). Standing on its base, it will tend to return to its former condition if disturbed, and so is *stable.* Lying on its side, it is in *neutral equilibrium* because it will remain in any position. When it is balanced on its point,

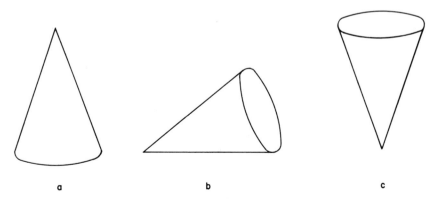

FIGURE 6.1 Cone in three states of equilibrium: (*a*) stable; (*b*) neutral; (*c*) unstable.

any disturbance will make it fall, a condition described as *unstable equilibrium.*

Consider a parcel of air, temperature 5°C (see Figure 6.2), which is lifted 1 km in the atmosphere. It will cool at the dry adiabatic lapse rate to a temperature of −5°C. If at its new level it is surrounded by air at a temperature of −3°C (Figure 6.2a), the lifted parcel is cooler than its surroundings, therefore more dense, and it will start to fall if released. If the lifted parcel of air is surrounded by air at a temperature of −5°C (Figure 6.2b), there will be no buoyancy forces acting on it, and it will tend to remain at the new level. In a third instance, if the surrounding air has a temperature of −8°C (Figure 6.2c), the rising parcel of air is warmer than its surroundings and, hence, buoyancy forces will cause it to continue to rise. Thus we have, respectively, stable equilibrium, neutral equilibrium, and unstable equilibrium in the air column.

Figure 6.3 presents the same results in graphical form. The coordinates are temperature and height, and AC gives the temperature change of a rising parcel of air. If at level z_2 the temperature surrounding the parcel is given by $B > C$, then the rising parcel will tend to subside back to A. If the environment curve is given by AD where $D < C$, the rising parcel is warmer than the environment, and the air column is unstable. Neutral stability occurs if the air column has a temperature given by the dry adiabatic lapse rate AC.

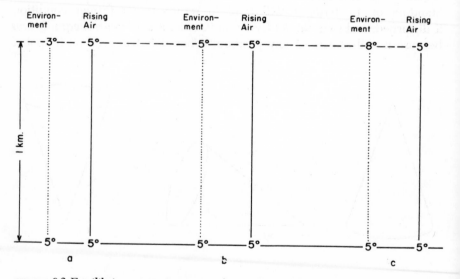

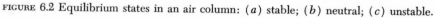

FIGURE 6.2 Equilibrium states in an air column: (*a*) stable; (*b*) neutral; (*c*) unstable.

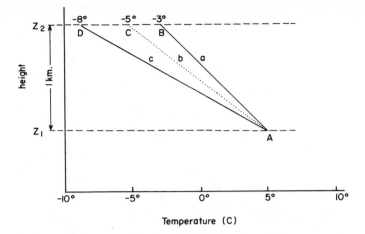

FIGURE 6.3 Stability on temperature-height graph: (a) stable lapse rate; (b) neutral lapse rate; (c) unstable lapse rate.

The discussion of the preceding paragraphs can be generalized. If dry air has an *environmental lapse rate, α,* equal to the dry adiabatic lapse rate, Γ, the air column is in neutral equilibrium. If the environmental lapse rate is less than the dry adiabatic lapse rate, the situation is stable. When the temperature of the air column changes with height at a rate greater than the dry adiabatic lapse rate, the condition is unstable. By using algebraic symbols, one can state that an air column is

$$\text{Stable when} \quad \alpha < \Gamma$$
$$\text{Neutral when} \quad \alpha = \Gamma$$
$$\text{Unstable when} \quad \alpha > \Gamma$$

In Section 2.4, an example was given of an air column where the temperature dropped $3°C$ in 16 m. This is at the rate of 19 degrees in 100 m, a value many times the dry adiabatic lapse rate, and so the situation was unstable. Under these circumstances, air parcels near the earth began and continued to rise, and air from aloft subsided to replace them. These formed, then, vertical eddies bringing cool air down from aloft and carrying warm air away from the ground. These eddies, with their different temperatures, caused the instantaneous temperatures to vary 1 to 2 deg on either side of the mean. The thermograph record given in Figure 2.10 also shows the effects of eddies formed during the afternoons. These eddies provide one of the ways by which the solar energy that is absorbed at the ground is carried aloft. Near the ground, the formation of eddies is inhibited, and extremely high lapse rates can occur. In the free atmosphere, vertical eddies readily form and adjust the temperature distribution to a near neutral condition.

Heating is extreme in desert areas where clear skies permit the solar radiation to warm the ground and little moisture is present to use up the energy in evaporation. In these regions, turbulent eddies form readily and to heights of 3 km or more. These vertical eddies form "thermals" or "air pockets" that cause bumpy riding in aircraft as they pass from an up-current to a down-current or vice versa.

The variation of both pressure and temperature poses a problem in determining which of two parcels of air has the more energy stored within it. This can be resolved by comparing the temperatures that would result if the air parcels were brought adiabatically to some standard level. The standard level is taken as 1000 mb, that is, very close to sea level. The temperature that an air parcel would have if carried dry adiabatically to 1000 mb is called its *potential temperature,* usually designated θ. Thus, air at 6 km, with $T = -35°C$, has $\theta = -35° + 6$ km $\times$ 10 deg km^{-1} = 25°C. It therefore has more stored energy than air at 1 km, $T = 5°C$, $\theta = 15°C$.

An air column with neutral stability has uniform potential temperature; θ remains constant. With stable conditions, the potential temperature increases with height; and under unstable conditions, the potential temperature decreases with height. In fact, the rate of change of potential temperature with height is a very simple method of determining the degree of stability or instability. One can say that the air is stable or unstable, depending on the sign of $\Delta\theta/\Delta z$. As will be recognized from the above discussion, the potential temperature is conservative to adiabatic changes.

6.3 *Stability of Saturated Air*

The presence of water presents an added complication to the discussion of stability. Rising air cools at very nearly the dry adiabatic lapse rate as long as it remains unsaturated. At some level, the air will reach its dew point, and then condensation will occur. The base of a cumulus cloud (see Figure 4.13) is found at this level. From this level upward, the cooling will cause moisture to condense as cloud droplets or ice crystals. This condensation releases the heat of evaporation that the vapor absorbed when it changed from a liquid. In Section 6.1, Equation 7

$$\Delta h = c_p \Delta T - \Delta p/\rho \qquad (9)$$

was treated with the heat Δh being considered as zero. With condensation, heat is being added from inside the parcel. If w_s is the mixing ratio of the parcel, Δw_s represents the moisture which condenses. Then $\Delta h = -L \, \Delta w_s$ where L is the latent heat. The negative sign is used because heat is added when the moisture decreases. The equation to represent adiabatic conditions when saturated air is being lifted is

$$-L \, \Delta w_s = c_p \, \Delta T - \Delta p / \rho \qquad (10)$$

A solution for this equation is beyond the scope of this book.

On the basis of Figure 4.2, we recognize that for high temperatures Δw_s is large for a given change of temperature, compared with the value when the temperature is low. Careful computations show that saturated air at 1000 mb and 30°C cools at the rate of 3.5 deg C km⁻¹; at 0°C, the rate is 6.6 deg C km⁻¹. Very cold air contains little moisture and so cools when rising at a rate only slightly less than the dry adiabatic lapse rate. With decreasing pressure, the lapse rate increases slowly. In this process the heat acquired by the water substance when it evaporated is being released again to become sensible heat and raise the temperature of the air. The resulting rate of decrease of temperature is called the *moist* or saturated *adiabatic lapse rate*, designated Γ_s.

When a current carries saturated air downward, compression will cause it to warm. If there is no water present in either liquid or solid form, the air will warm at the dry adiabatic lapse rate of 10 deg km⁻¹. Usually it subsides inside the cloud where it can absorb moisture from the cloud droplets. The evaporation cools the air so that it remains saturated, and the air heats with subsidence at the same rate it cooled (the moist adiabatic lapse rate). Observations within cumulus clouds, with their updrafts and downdrafts, show that the drop in temperature is close to the moist adiabatic, although some mixing with the cooler, drier air outside the cloud changes the lapse rate slightly.

By recognizing that saturated air cools at the moist adiabatic lapse rate, we can again go through the argument of Section 6.2 on stability to show that for rising saturated air, the air column

is stable if $\qquad\qquad \alpha < \Gamma_s$
has neutral stability if $\alpha = \Gamma_s$
is unstable if $\qquad\qquad \alpha > \Gamma_s$

Care must be taken in considering subsiding air (see Exercise 9 at the end of the chapter) for the rate of warming depends on the presence or absence of moisture, and whether evaporation continues to keep the subsiding air saturated.

Aircraft observations have confirmed that the air temperature within a cumulus cloud, that is, in the rising air, is higher than the one surrounding the cloud at the same level. It is this temperature difference that provides the buoyancy force to drive the air currents upward.

Notice that the water condensing releases the latent heat that was used when it evaporated at the surface of the earth. This additional heat, being supplied at the base of the air column, changes the temperature difference between the cloud and the surrounding air. When the change is such that the parcel of air in the cloud is warmer than the one outside, buoyancy forces drive the cloud parcel upward. The air has become unstable and vertical currents form to reduce the instability.

6.4 *Thermodynamic Diagrams for Upper-air Study*

Sections 6.2 and 6.3 have shown that the vertical distribution of temperature in an air column determines the degree of stability of the air and, therefore, the tendency for vertical motion. It is therefore desirable to examine temperature distributions rapidly, which is most easily done from suitable graphs.

The most logical graph, it would seem, would be one where the temperature is plotted against height. This is sometimes done, as in Figures 2.14 and 6.3. A graph of this kind ignores the fact that meteorological observations record pressure, not height, and that meteorologists prefer to use pressure coordinates rather than height coordinates in many of their relationships. A more practical diagram makes pressure the variable for the vertical coordinate. When pressure is plotted on a logarithmic scale, the result is not far different from height plotted on a linear scale. Such a diagram is the *emagram* (Figure 6.4).

The preceding two sections suggest two families of curves that are useful in determining stability. They are the dry adiabats, or lines of equal potential temperature, and the moist adiabats, or the paths that saturated parcels of air follow when lifted in the atmosphere. In the emagram, the dry adiabats are straight lines rising from right to left (the heavy lines in Figure 6.4). The moist adiabats are shown on the emagram as curves like the dashed lines in Figure 6.4, rising rapidly at the warm side of the diagram but becoming almost parallel to the dry adiabats with low temperatures.

A fifth set of lines is shown by dots in Figure 6.4. They are lines giving the amount of moisture in grams per kilogram required to saturate the air. The amount of moisture required to saturate a given space is determined solely by the temperature. But the saturation mixing ratio, which divides the mass of moisture by the mass of air, increases for a given temperature as the pressure falls (see Exercise 13, Chapter 4). Therefore, the lines of constant mixing ratio intersect the isotherms at a small angle.

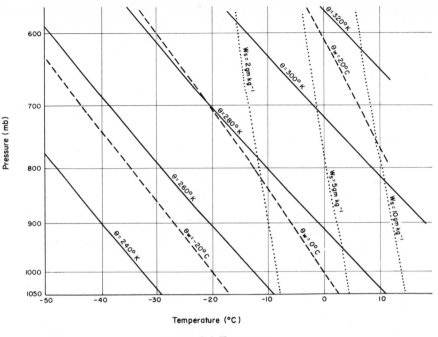

FIGURE 6.4 Emagram.

Another common diagram for plotting upper-air temperatures is the *tephigram.* Like the emagram, it has five basic sets of curves. In it, the adiabats and the isotherms are perpendicular (Figure 6.5). If the chart is rotated so that the adiabats are horizontal and the isotherms vertical, the isobars rise from left to right, intersecting the other two sets of lines at an angle of approximately 45°. The exact angle depends on the relative scales of θ and T. The moist adiabats again curve upward from right to left, becoming nearly parallel to the dry adiabats at temperatures below —35°C. The mixing ratio lines, as in the emagram, intersect the isotherms at a small angle.

The emagram and tephigram are both used in upper-air analyses. The tephigram is less easily comprehended but has one compensating advantage. The temperature in the free atmosphere usually drops at a rate somewhere between isothermal and the dry adiabatic. These lapse rates are, in the tephigram, distributed over a right angle rather than over an angle of 45° as in the emagram. Therefore, significant differences in lapse rates are more readily distinguished on the tephigram.

The tephigram or emagram is useful in many problems dealing with the atmosphere. Consider the problems relating to measurements of

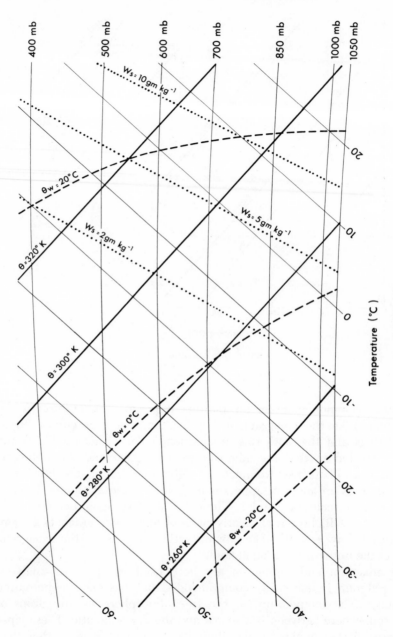

FIGURE 6.5 Tephigram.

124

moisture in Chapter 4. If at constant pressure the temperature and dew point are known, the mixing ratio lines give the actual moisture as the amount required to saturate the air at the dew point and, also, the saturation mixing ratio at the actual temperature. From them, the relative humidity can be readily determined.

The thermodynamic diagrams are even more useful in considering the ascents of rising parcels of air. Consider a parcel of air, $p = 900$ mb, $T = 18°C$, $T_D = 7°C$ (Figure 6.6). According to the tephigram, $\theta = 300°K$, $w = 7$ gm kg^{-1}, $w_s = 14.7$ gm kg^{-1}, and so the relative humidity is 48 per cent. If the air rises to 850 mb, the path on the tephigram is given by the line $\theta = 300°K$, and it reaches 850 mb at a temperature of 13.5°C. The moisture is still 7 gm kg^{-1} and, hence, the dew point has dropped to 6°C, and the relative humidity risen to 60 per cent.

With further lifting, the relative humidity increases to become 100 per cent at 760 mb. This is shown on the tephigram as A, the point of intersection of the lines $\theta = 300°K$ and $w = 7$ gm kg^{-1}. Because any further lifting will cause water vapor to condense and a cloud to form, this level is called the *lifting condensation level*. The rate of decrease is now given by the moist adiabat AB. If the parcel is lifted to B at 500 mb, the temperature will decrease to $-16°C$, and the mixing ratio decrease to 2.3 gm kg^{-1}. The lost 4.7 gm has condensed into water

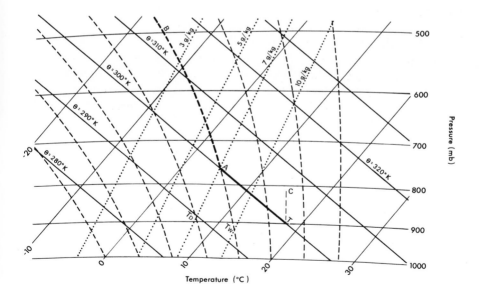

FIGURE 6.6 Path of a parcel of moist air, $p = 900$ mb, $T = 18°C$, $T_D = 7°C$, as plotted on a tephigram.

droplets or ice crystals. Some of the condensed moisture may fall out of the cloud as snow or rain.

As discussed in Section 6.3, the path on the tephigram or emagram that is followed when air subsides is not unique but depends somewhat on the conditions imposed. If the air subsides within the cloud and the moisture reevaporates until it is all gone, the path on the tephigram will be the reverse of the upward path.

6.5 *Wet-bulb and Wet-bulb Potential Temperatures*

Consider once again the parcel of air, illustrated in Figure 6.6. It cools at the dry adiabatic lapse rate until it reaches the lifting condensation level at 760 mb, and at the moist adiabatic lapse rate above that level. With subsidence in the same environment, the air would return to the original pressure (900 mb) and to the same temperature. The moisture which condensed above 760 mb would reevaporate as the parcel subsided until, at the lifting condensation level, it has all changed to vapor once again.

If the environment were changed so that below 760 mb there was still liquid moisture, the subsiding air parcel would continue along the moist adiabat AT_w reaching 900 mb with a temperature of 11.5°C. Compare this process with the determination of the wet-bulb temperature as given in Section 4.3. In both processes, we begin with partly saturated air and, with no addition or subtraction of heat, finish with saturated air at the same pressure. The tephigram, then, provides a method for determining the wet-bulb temperature of the air. Notice the relationship among the three temperature lines: *the dry adiabat through the temperature, the moist adiabat through the wet-bulb temperature, and the mixing ratio line through the dew point intersect at the lifting condensation level.* The knowledge of any two temperatures permits the determination of the third. Other data, such as the mixing ratio or the relative humidity, may also be used as the initial information.

The energy contained in any parcel of air is partly sensible heat, measured by the temperature, and partly the latent heat of the water vapor present. The change by which the wet-bulb temperature is reached does not change the total amount of energy, although it changes the separate parts. Also, at the wet-bulb temperature the amount of moisture is uniquely determined. Therefore, the wet-bulb temperature is a measure of the total energy of the air parcel, and the energy of two parcels may be compared by means of their wet-bulb temperatures. The comparison of two air masses at two different pressures is done in the same manner as with potential temperatures. The same standard pressure (1000 mb) is specified, and comparable values are obtained by bringing the wet-bulb temperature along the moist adiabat to 1000 mb to give the *wet-bulb potential temperature*, usually designated by θ_w. In the example of Figure 6.6, $\theta_w = 16°C$. Each moist adiabat may be identified by the wet-bulb potential temperature associated with it.

6.6 Stability for Moist Air

The stability of unsaturated air can best be examined with the help of a thermodynamic diagram. If the air at 900 mb in Figure 6.6 is lifted, it follows the path TAB. This must be compared with the environment curve, that is, the temperature-height curve for the air column in which the parcel is rising. If the environment lapse rate α is greater than Γ, the environment curve will lie below TA on the tephigram. In this case, the rising parcel of air is always warmer than the surrounding air, and so the situation is unstable. If, on the other hand $\alpha < \Gamma_s$, in this instance less than 5 deg km^{-1}, the rising air is always colder than the environment, and, thus, the air parcel is stable. In this case, the environment curve on the tephigram will lie to the right of TC, the moist adiabat through T.

If the environment curve lies between TA and TC, it may and may not intersect the curve AB. If the two curves intersect at D (not shown on the diagram), the rising parcel will be stable unless it is lifted to D, and unstable if it goes above D. The point D depends on the value of α and the humidity of the air, which determine the point A. This type of instability is called *conditional instability*, and the level D the *level of free convection*.

An example of an environment curve where the lapse rate was between the dry and moist adiabats is given in Figure 6.7. This gives the temperature ascent for Oklahoma City for 1800 h CST 5 June 1962, an afternoon during which ten tornadoes were reported in the state of Oklahoma. Significant points on the ascent, for those who wish to obtain a larger scale diagram on their own charts, are:

P	966	850	815	533	243	170	150	mb
T	29.4	19.2	19.2	−8.0	−49.8	−65.4	−63.9	°C
T_D	21	14	5	−25	−	−	−	°C

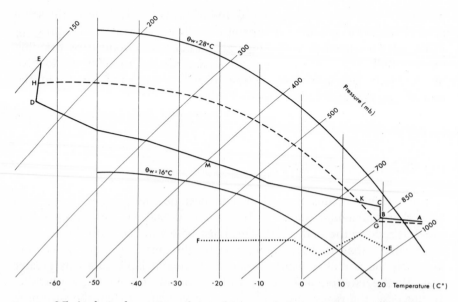

FIGURE 6.7 Analysis, by means of a tephigram, of the air column over Oklahoma City, 1800 h CST 5 June 1962.

The lapse rate from the surface to 850 mb, *AB*, was nearly adiabatic, a result of surface heating during the afternoon. Above an isothermal layer *BC*, the air was much drier but had a high lapse rate to the top of the troposphere at 170 mb, *D*. The dotted irregular line *EF* gives the dew point curve for the ascent.

When the air parcel at *A* was lifted, it reached the lifting condensation level at *G*, 850 mb, slightly colder than the environment. Further lifting caused it to cool along a moist adiabat *GH*. It remained colder than the environment until the level of free convection at 760 mb, where the moist adiabat *GH* intersects the environment curve *CD* at *K*. From 760 mb the air parcel rose by buoyancy forces until the curve again intersects the environment curve. In this instance, the intersection was at *H*, 160 mb, in the stratosphere.

The tephigram is designed in such a manner that an area is proportional to energy. Thus the area between the two curves *ABCK* and *AGK* represents the energy required from outside sources, such as mechanical energy, to lift the parcel at *A* to the level of free convection. In the example this is relatively small compared to the energy represented by the area between the environment curve *KDH* and the moist adiabat KH. Thus a rising parcel of air would have been able to acquire considerable energy from the unstable environment. Some of this energy was released as kinetic energy in the development of the tornadoes in the area during the afternoon. If, as is usually the case, the energy available for release, $KDHK$ in the diagram, is much less than the energy necessary to carry the parcel aloft, $ABCKGA$, then the vertical currents will be small or nonexistent.

In the example of Figure 6.7 the cloud base was at the lifting condensation level, 850 mb or approximately 1 km above the surface. After the parcel passed the level of free convection, buoyancy forces supplied energy to cause it to acquire high vertical velocities. When it reached the level of H, in the stratosphere, the buoyancy forces were lacking Instead, in the stable layer in the stratosphere, the vertical motion was halted or was diverted to a horizontal motion. Thus, H gives approximately the cloud top, and the horizontal motion provided an anvil such as is often seen at the top of a cumulonimbus cloud (Figure 4.14). In this instance, the cloud had a depth exceeding 12 km.

The point M at 400 mb in Figure 6.7, where the environment curve is tangent to a moist adiabat, is significant. If the path of the rising parcel is to the left of M, the parcel will always remain stable. By following down the moist adiabat through M to the surface pressure, one can determine that this situation would occur if the surface wet-bulb temperature were below 17°C. This is another method of stating the "condition" for conditional instability. Under some circumstances, the surface wet-bulb temperature will pass through the critical value as the heating of the day causes the surface temperature to rise rapidly and the moisture to increase slowly. It is unlikely that cumulus clouds will develop until this time is reached. The rapidity with which the development occurs after the initial formation and the level to which the currents rise depend on the lapse rate above the level of free convection.

Figure 4.15a shows a picture of a cumulus cloud, taken at Penhold, Alberta at 1925 h MST 15 July 1968. The temperature and moisture distributions in the air column at Penhold at 1700 h MST are given in Figure 6.8. The surface layer showed absolute instability with conditional instability aloft. A tephigram analysis shows that the lifting condensation level of the surface air was 770 mb, and that, above the LCL, the rising curve would pass through the tropopause at 300 mb, and cross the environment curve at approximately 265 mb, in close agreement with the cloud top as determined by radar (see Section 4.7). It is interesting to use a thermodynamic chart in a situation like the one illustrated

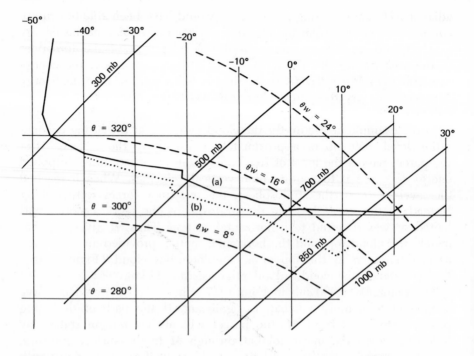

FIGURE 6.8 Temperature (*a*) and dew point (*b*) above Penhold, Alberta, 1700 h MST 15 July 1968.

in Figure 6.8 to observe the change that may occur with only a few degrees difference in the temperature of the surface air.

6.7 *Stability Considerations with Movements of Layers*

The discussion in the preceding sections has considered the effects on a parcel of air detached from its environment and caused to move upward or downward. At times the air over a wide area will move vertically. It may move as a unit, or there may be a flow inward so that the horizontal dimensions become less and there is a stretching in the vertical. The reverse also occurs, that is, a spreading out horizontally and contraction vertically. These processes are described by the terms *convergence* and *divergence*. In the second situation, there is horizontal divergence and vertical convergence.

These mass movements cause changes in the degrees of stability. First, consider dry air. Usually the air is absolutely stable, which can be expressed by stating that the potential temperature θ increases with height, that is,

$$\frac{\Delta\theta}{\Delta z} > 0$$

Because for each parcel, θ is constant with vertical movements, $\Delta\theta$ for a layer is also constant, but Δz probably changes. If there is no convergence or divergence, the value of Δp, the change in pressure from the bottom to the top of the layer, is constant. With subsidence and compression, Δz becomes less and so $\Delta\theta/\Delta z$ becomes greater, and the air becomes more stable. With horizontal divergence, usually found in the centers of anticyclones, the decrease of Δz is even greater and, hence, the air becomes even more stable. With rising air, particularly if accompanied by horizontal convergence, the value of Δz increases and so the air becomes less stable. Yet under these conditions, $\Delta\theta$ must still remain positive and, consequently, absolute instability cannot develop.

With saturated air, the situation is similar. If the air is initially stable, vertical movement may change the degree of stability but will not produce instability. Unstable air may become more or less unstable, but cannot become stable if the air remains saturated.

The situation is less simple with moist air. As long as the rising layer remains unsaturated, the situation is the same as with dry air. If the relative humidity is approximately constant with height and if $\Gamma_s < \alpha < \Gamma$, the layer will become saturated with lifting with little change in α. Nevertheless, the layer will now have absolute instability because the environmental lapse rate is being compared with Γ_s, a smaller quantity.

To understand the situation when the relative humidity is not constant, it is simplest to refer to an upper-air chart. Consider a layer between 980 mb, $T = 5°C$, $T_D = 4°C$, and 880 mb, $T = 2°C$, $T_D = -5°C$ (Figure 6.9). This is the type of air that could be brought in winter by west winds to the Washington coast. The surface layer has increased its moisture through evaporation, but in the stable situation this moisture stays near the surface.

In Figure 6.9, AB is the temperature curve and CD the dew point curve. Winds will force this layer to rise over the coastal mountains; assume it rises 100 mb without divergence. With this lift, the total layer will be saturated, and the lifted layer will be represented by EF. The air at A will cool at rate Γ for only 15 mb lift, after which it will cool at rate Γ_s. The air at B will cool at rate Γ for 90 mb, and at rate Γ_s for only 10 mb. Therefore, the lapse rate α has increased.

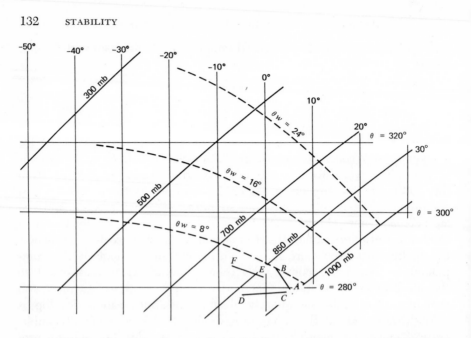

FIGURE 6.9 Development of convective instability as shown on a tephigram. *AB* gives initial temperature, *CD* initial dew point, and *EF* temperature after 100 mb lift.

Also α is to be compared with Γ_s. The tephigram shows that the final value of α is slightly greater than Γ_s. The example shows that, under certain circumstances, unstable air may develop through the lifting of air that is *convectively unstable*. In convectively unstable air, the wet-bulb potential temperature decreases with height. These circumstances are not uncommon during the winter along the Pacific coast from northern California to Alaska and give rise to the winter thunderstorms over the coastal mountains.

6.8 *Radiation and Stability*

As described in Chapter 5, the greatest changes of temperature occur at the earth's surface. These changes affect the air above the ground, particularly by radiation. At night, the exchange of long-wave radiation cools the surface layers most, so that an inversion develops above the earth's surface. The depth of the inversion depends on the dryness of the air because water vapor will absorb and reradiate terrestrial radiation.

During the day the ground absorbs solar heat, to raise its temperature above the air temperature. Once again, the effects of radiation on the air are greatest just above the surface. The column becomes unstable, and vertical currents are set up to transfer the heat from the earth to higher levels. These currents mix the lowest layers and tend to produce a dry adiabatic lapse rate below the lifting condensation level.

When a cloud deck covers the earth, radiant energy is exchanged between the cloud base and the earth's surface. Normally, the exchange gives a net positive balance to the cloud base, warming it. At the top of the cloud the radiative exchange results in a net loss because the solar radiation and long-wave radiation from the clear air above warm the cloud only slightly, and the top of the cloud radiates heat in all wavelengths.

With warming at the base and cooling at the top, a cloud can rapidly develop an unstable lapse rate even when originally the lapse rate within the cloud was stable. The vertical currents that are produced within the cloud cause irregularities in the cloud tops. Aviators are well aware of the fact that the top of a cloud layer does not have the same uniform appearance as the cloud base. With further development, the vertical currents at times become apparent in the base of a stratus cloud, causing varying shades of grey or even holes in the overcast deck. Thus, radiation differences may change a stratus cloud into stratocumulus.

6.9 *Inversions*

In the preceding sections of the chapter we have been concerned with unstable conditions. They arise because the earth's surface is the major source of heat for the atmosphere. But the surface can cool the lowest layers, to produce a stable condition. The cooling may arise when advection moves air from a warm region to a cold as, for example, from heated land to a cool lake surface. An inversion frequently develops at night when the solar heat is lacking and the earth cools by long-wave radiation. An inversion, whatever its cause, has its own peculiar weather phenomena.

With an inversion, the movement of an air parcel away from its original level is inhibited by buoyancy forces. A ribbon of smoke from a fire rises in an inversion until, through cooling by radiation and mixing, it reaches a level where it is in equilibrium with its environment. It then moves horizontally with little lateral spreading. An immediate result of this phenomenon is that the pollutants which mankind ejects into the lowest layers remain within these layers. The light winds (see Section 7.2) transport them slowly, permitting the concentration of the noxious gases to increase.

The urban environment provides an additional meteorological phenomenon. The central area is frequently warmer at night than the surroundings (see Figure 2.6), so that, over this area, there can be an unstable layer for a few hundred feet topped by the inversion. In the unstable layer, vertical currents and mixing produce a uniform distribution of pollutants throughout the layer.

A similar situation occurs at times in open country in the vicinity of a high smokestack. During the night, the gases from the stack tend to remain at the level where they are in equilibrium. After sunrise, the surface temperature rises and a layer develops where the lapse rate is nearly adiabatic. When the top of this layer reaches the level at which the gases are concentrated, the mixing brings them to the ground. Thus we tend to have a maximum of pollution in the morning. In pollution studies, this process is called *fumigation*. Later with increasing temperature, the mixing distributes the gas fumes over a thicker layer and the ground concentration decreases.

PROBLEMS AND EXERCISES

A thermodynamic diagram* is necessary for some of the following problems.

1. Given certain measures among the following, determine the others: temperature (T), dew point (T_D), mixing ratio (w), wet-bulb temperature (T_w), lifting condensation level (LCL), relative humidity (RH), potential temperature (θ), wet-bulb potential temperature (θ_w).

(a) $p = 1000$ mb, $T = 15.7°C$, $T_D = 12.2°C$
(b) $p = 920$ mb, $T = 6.3°C$, $T_w = 1.0°C$
(c) $p = 800$ mb, $\theta = 290°K$, $w = 3.0$ gm kg^{-1}
(d) $p = 1032$ mb, $T = -25°C$, $RH = 75$ per cent
(e) $p = 950$ mb, $LCL = 765$ mb, $w = 8.0$ gm kg^{-1}

2. Air, $p = 1020$ mb, $T = 22°C$, $T_D = 11°C$, is lifted adiabatically to 400 mb. How much water will condense from the air? If the condensed water forms water droplets of radius 10^{-2} cm, how many droplets will there be in 1 m^3?

3. Air is at a pressure of 1000 mb, temperature 18°C, and mixing

---

* Thermodynamic diagrams, tephigrams or emagrams, are used by national meteorological services, from which copies may usually be obtained.

ratio is 12 gm kg⁻¹. Assume that the water vapor condenses and the heat is used to warm the air. Determine the resulting temperature. Check your result with the value obtained if the air is lifted adiabatically to the top of the atmosphere and then brought back dry adiabatically to 1000 mb. (This value is called the *equivalent temperature.*)

4. Plot the following five radiosonde ascents:

(a) Tatoosh, Washington
1200 GMT
3 February 1962

p	T	T_D
1014	7.8	7.8
860	1.8	−2
818	1.8	−18
726	−3.0	−10
692	−4.5	−22
400	−34.2	
337	−39.8	
225	−49.6	
206	−56.0	
153	−45.3	

(b) Montgomery, Alabama
0000 GMT
13 June 1961

p	T	T_D
1002	27.2	22
850	17.1	12
700	6.7	3
635	1.2	−1
506	−7.0	−20
400	−17.7	−27
300	−34.3	
200	−55.9	
169	−64.3	
150	−63.0	
116	−70.0	
100	−69.1	

(c) Tucson, Arizona
0000 GMT
13 June 1961

p	T	T_D
919	35.4	−4
850	28.3	−8
670	7.4	−17
653	7.3	−17
500	−7.3	−29
400	−19.2	−37
244	−47.9	
200	−56.7	
174	−60.9	
150	−59.7	
100	−67.0	
89	−67.4	

(d) Inuvik, Northwest Territories
1200 GMT
17 December 1961

p	T	T_D
1024	−44.5	
1010	−32.6	−37
930	−30.3	−37
904	−28.0	−34
797	−29.9	−35
600	−35.8	−41
536	−40.0	−44
326	−59.6	
250	−59.4	
200	−58.2	
150	−58.4	
137	−58.4	
102	−61.8	

(e) Alert, Northwest Territories
1200 GMT
18 July 1962

p	T	T_D
1015	1.6	1
995	−0.3	−3
981	2.5	0
927	1.0	−1
877	5.1	1
833	2.7	−2
650	−10.9	−12
556	−15.8	−32
469	−25.5	−40
400	−32.8	
300	−49.1	
260	−53.0	
250	−51.7	
220	−47.6	
200	−46.8	
150	−45.0	
100	−45.1	

(1) List the tropopause for each ascent.
(2) List the inversions in the troposphere.
(3) List any absolutely unstable layers.
(4) Examine the points below 850 mb for conditional instability.
(5) What would be the cloud base and cloud top for a cumulo-formed cloud at Montgomery?
(6) Lift the layer from the surface to 818 mb for Tatoosh 100 mb, and examine the resultant curve for stability.

5. Air is lifted dry adiabatically from a pressure of 1000 mb, temperature 27°C, to 800 mb. By using Equation 7, Section 6.1, compute approximately the resulting temperature. Check your answer by means of a thermodynamic diagram.

6. A mass of air at 850 mb has a temperature of 10.0°C and dew point of 2.8°C. Find the temperature if this air is lifted adiabatically to 500 mb. If the air at 500 mb has a temperature of −17.7°C, find the difference between the temperatures of the environment and the rising air. (This difference is called the *Showalter stability index*.)

7. Air with a temperature of 20°C, pressure of 1000 mb, and mixing ratio 9 gm kg⁻¹, is lifted adiabatically to 700 mb. Determine the final

conditions. Insert values from your results into Equation 10, Section 6.3, and compare the two sides of your equation. (The two sides will not be exactly equal, but the error should be small.)

8. In an early morning ascent at Omaha, Nebraska, the following observations were made:

Pressure	966	955	946	850	700	651	640	500	400	mb
Temperature	15.9	21.0	23.4	16.9	3.6	0.0	1.9	−10.7	−23.8	°C
Mixing Ratio	10.8	11.3	8.5	5.4	4.2	2.5	2.0	0.9	0.2	gm kg⁻¹
		300	250	233	200	mb				
		−39.7	−49.3	−51.9	−52.5	°C				

Consider the change in stability as the surface temperature rises during the morning to 20°C, 25°C, 26°C. What will be the cloud bases and tops under these conditions?

9. Air at 950 mb, $T = 24°C$, $T_D = 10°C$, is lifted to 350 mb. Three quarters of the condensed water falls as rain. Then the air subsides back to its initial pressure. Determine the resulting temperature.

10. In Section 6.2, stability conditions were discussed by considering a parcel of air rising 1 km. Show that the same conditions hold true if the air is caused to subside 1 km.

11. Air with $p = 950$ mb, $T = 80°F$, $T_D = 65°F$, is carried aloft to 400 mb. Four fifths of the condensed water falls as rain. If 2.5 cm falls in 1 hr, how much air has been carried aloft? Given that the mass which moves through a unit cross section in unit time equals the velocity times the density, compute the vertical velocity at 400 mb for this rainfall.

CHAPTER 7

WINDS AND PRESSURE

7.1 Surface Wind and Its Measurement

Everybody knows about the wind and its effect. On a hot day the wind makes the weather more bearable; on a cold day, more severe. It blows our sailing ships and our windmills, carries away and dilutes our smoke and sprays, and makes waves on our lakes and oceans. When it is strong, it can cause widespread damage. It is proverbially variable, changing rapidly both its direction and speed. Yet, as described in Chapter 1, the winds of the earth are highly significant in the distribution of solar energy and in the balancing of the heat budget of the earth. It becomes necessary therefore to understand something of the basic laws governing these movements of air.

Reference to the various wind directions and speeds is found in the earliest literature. The Greeks had a "Tower of the Winds," and names for the different directions. Fishermen of the high seas had descriptive names for the different wind velocities. Admiral Beaufort, in the early part of the nineteenth century, made the terms more definite by dividing wind speeds into 12 classes, using some of the mariners' terms for describing the classes. Each class was defined in terms of the effects of the wind on waves. Thus a "strong breeze" causes 10- to 12-foot waves with streaks of spray from foamy crests. Beaufort's scale of winds was later described in terms of the effects on land. Thus, on land, a strong breeze puts large branches in motion, and causes whistling in telegraph wires.

The development of instruments to measure and record wind direction and speed has caused Beaufort's classes to be replaced by numerical values of the wind speeds. Yet remnants of Beaufort's scale remain. Thus a tropical storm becomes a hurricane only when the maximum wind speed reaches "hurricane force," that is, 75 miles per hour. Surface winds are still reported by the points of the compass from which they blow, usually to eight points, but at times to 16 points. Wind directions in the free atmosphere are usually measured and reported in degrees,

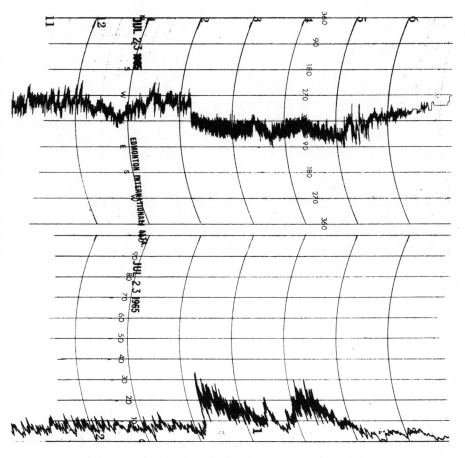

FIGURE 7.1 Wind speed (mi hr⁻¹) and direction, as given by a Dines anemometer, Edmonton International Airport, Alberta, 1100 h to 1800 h MST 23 July 1965.

measured clockwise from north. Thus an east wind is reported as 90°; south, 180°; west, 270°; and north, 360°.

The variability of the wind still presents problems. Figure 7.1 gives a record of the instantaneous wind direction and speed for Edmonton International Airport for the afternoon of 23 July 1965. Until 2 P.M., the wind was generally from the northwest quadrant with a speed under 10 mi hr^{-1}. The actual direction varied as much as 90° during most 10-minute periods. Shortly after 2 P.M., the wind suddenly shifted to east of north and the speed increased to 20 mi hr^{-1} with some gusts over 30 mi hr^{-1}. After 5 P.M. the wind died down and again shifted to northwest. A *mean wind speed* and *direction* can be determined under these circumstances, but any user should recognize that the instantaneous value can vary widely from the mean. The record in Figure 7.1 came from a *Dines pressure-tube anemometer* which is able to measure rapid fluctuations in the wind.

A more practical method of obtaining the mean wind speed comes from the *three-cup anemometer* (Figure 7.2). The number of rotations of the cups is counted and the mean wind determined. A wind vane is usually attached to the anemometer to record the wind direction.

Following the standard terminology of physicists and meteorologists, the wind *velocity* gives both the speed and direction of the wind. The wind *speed* gives the rate of motion but does not specify the direction.

7.2 Distribution of Winds Near the Earth's Surface

In the preceding section, we pointed out that the wind can vary considerably in a short time. The variation in space has been noted by all. For instance, wind speed and direction can be considerably different to the windward, at the side, and to the leeward of a large building. The effects of a grove of trees are equally well marked. The contours of the earth also cause changes in the wind velocity. For example, a valley tends to funnel the winds along its direction.

Even over smooth level land, the wind speed changes with height. Figure 7.3 presents the change of wind below 16 m under different circumstances. Notice that height is given on a logarithmic scale. Above 5 m the wind increases with height but, in general, slowly. Below 5 m, and particularly in the lowest 2 m, the change of wind with height can be considerable. Over rough terrain, the changes are more rapid and more irregular.

The diurnal rhythm of temperature produces a corresponding rhythm in the wind. Above a boundary layer of approximately $\frac{1}{2}$ km, the winds

FIGURE 7.2 Three-cup anemometer. (Courtesy, Meteorology Research Associates.)

are driven by forces, to be discussed in later sections, which have no diurnal rhythm. Energy from these levels is transferred downward, thus keeping surface winds blowing, winds which otherwise would die out because of friction. In unstable conditions, the vertical eddies that transfer this energy downward are large and vigorous. Surface winds then tend to differ only slightly from the upper winds. With inversions the eddies are small and the transfer of energy is slow. Surface winds tend to die down, and the change of wind with height is rapid.

To illustrate these effects, Figure 7.4 gives the mean hourly wind speeds at 13.5 m for seven winter months of 1966 to 1968 and six summer months of the same years for the Industrial Airport at Edmonton, Alberta, and at 63 m on the top of a building three miles from the airport. An inversion is present over the area most of the winter, with the January mean temperature at 850 mb (730 m above the surface) 4°C warmer than the surface temperature. The slight diurnal heating of winter is accompanied by a slight increase in the airport wind as shown in curve (a). Curve (c), the winter curve for 63 m, shows that the energy for this increase came from layers near 100 m and caused a reduction in wind there. The wind increase during the daytime at the airport in July [curve (b)] was much greater than the one in January. Curve (d) for 63 m is interesting. Durink the night, the winds are definitely

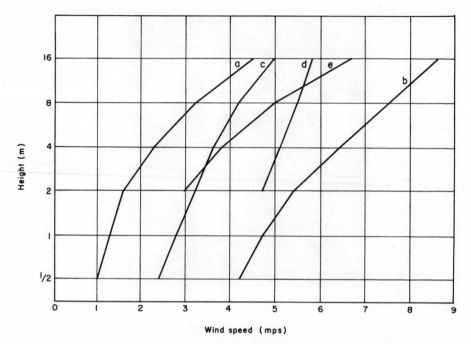

FIGURE 7.3 Wind speeds near the earth's surface, plotted on semilogarithmic paper: (*a*) and (*b*) over prairie grass, Suffield, Alberta; (*c*) over snow, Suffield, Alberta; (*d*) and (*e*) over clipped grass, O'Neill, Nebraska.

higher than at the ground. During the fumigation period at 07-09 h (Section 6.9), energy passes from levels above 50 m to the surface, and the winds at 63 m decrease. As the temperature increases, the adiabatic layer deepens. The energy to counteract the friction now comes from higher levels, and winds at both observation points increase to give approximately the same speed until 18 h. In the late afternoon both winds die down. But as the inversion increases in intensity the surface winds become light, while the winds at 63 m increase to their nighttime value.

Over a relatively smooth surface, the variation of the wind speed v with height near the surface is given approximately by the relationship

$$v = Kz^n$$

where z is the height and K is a constant determined by the initial conditions. In this equation the exponent n varies with stability, being approximately 1/7 under neutral conditions. If from Figure 7.3d we take $v = 5.8$ m sec^{-1} at $z = 16$ m, then $K = 3.9$ m$^{\frac{6}{7}}$ sec^{-1}. With this value,

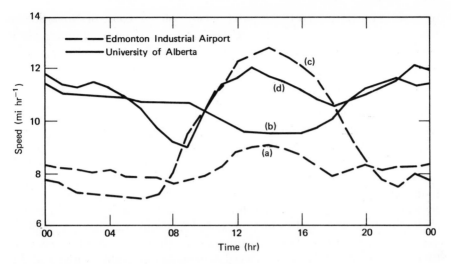

FIGURE 7.4 Mean hourly wind speeds: Edmonton Industrial Airport (13.5 m), (a) winter, (b) summer; University of Alberta, Edmonton (63 m), (c) winter, (d) summer.

the wind speed at 4 m is computed to be 4.8 m sec^{-1}, agreeing well with the observed value of 5.1 m sec^{-1}. For 2 m, the computed and observed values were 4.3 and 4.8 m sec^{-1}, respectively. The other profiles were not taken under neutral conditions, and other values of n must be used.

Wind reports are kept by direction and speed. They can be summarized by giving the prevailing wind direction and the mean speed regardless of wind direction. A graphical method of presenting more detail of the wind is given by a wind rose, illustrated in Figure 7.5. Shafts from a central point have lengths that are proportional to the frequency of the winds from that direction. Figure 7.5 gives the wind

FIGURE 7.5 Wind roses for Sable Island, Nova Scotia for (a) January, (b) July.

roses for January and July for Sable Island, off the coast of Nova Scotia. The difference shows the change between winter and summer winds over the northeastern United States and eastern Canada.

7.3 *Winds in the Free Atmosphere*

The determination of winds above the earth is made by releasing balloons and following their paths. At first the balloons were followed by means of a theodolite. This method is of limited value. The balloon can be followed only until it enters a cloud and, consequently, winds cannot be observed during periods of bad weather. More recently, the best observations come from tracking the radiosonde balloons either by radar or by a radio direction finder. Winds can be observed in almost all kinds of weather and to high altitudes. The pressure and temperature observations permit a determination of the altitude of the balloon and, therefore, the level of the observed winds. Regular observations at 12-hour intervals are taken by this method at many stations in all parts of the world. As pointed out in Section 7.1, the wind direction is given in degrees from north, instead of by compass points.

A wind observed in the free atmosphere is in reality a mean wind for a layer, because the value is determined by the path of a balloon carried in the wind stream. Observations on smoke puffs, etc., show that there are minor variations in the wind aloft, but no analysis of instantaneous velocities corresponding to an analysis of surface winds is possible.

7.4 *Pressure Gradient Force*

To understand the winds, we must seek the forces that keep the air in motion. If there were no such forces, the friction among the molecules would bring the air gradually to a standstill. The main cause of the wind is the *pressure gradient force* which is derived from the horizontal pressure gradient. We observed in Section 3.6 that determination of this latter quantity was the reason for the computation of sea-level pressures. Consider a volume of unit cross section with length Δx parallel to the x-axis (Figure 7.6). Let p be the pressure at one end, and $p + \Delta p$ the pressure at the other. By the definition of pressure, they give forces acting on the volume. The net force Δp acts to move the particles from high pressure toward low. The mass of air is $\rho \Delta x$ and, hence, the resultant force per unit mass is

$$F_x = -\frac{1}{\rho}\frac{\Delta p}{\Delta x}$$

(1)

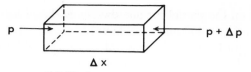

p ──── ← $p + \Delta p$

Δx

FIGURE 7.6 Horizontal pressure gradient force.

The negative sign is attached since the force acts in the direction of decreasing pressure.

A similar force, but acting vertically, is found for the hydrostatic equation (Section 3.4). In this latter equation, the force is balanced by the force of gravity. This is the usual situation and, as a result, vertical motions are small except when influenced by topography. When the air is unstable, there is no longer a balance and vertical motions develop (see Chapter 6).

The horizontal pressure gradient force of Equation 1 can act in all fluids. It is the cause of the flow in pipes, and can initiate horizontal movement in a pond with temperature differences in the water. In the atmosphere, if no other forces were acting, the pressure gradient force would cause all particles to move toward the points of lowest pressure. The result would be an equalization of pressure in the horizontal, the pressure gradient force would drop to zero, and the atmosphere would come to rest. These other forces are discussed in Sections 7.5, 7.7, and 7.8.

The weather map for 8 January 1969 (Figure 3.8) shows that the pressure dropped 21 mb between Indianapolis, Indiana and Kansas City, Missouri, 700 km distant. Let us compute the pressure gradient force and the motion of a stationary parcel if no other forces were present. With $p = 1010$ mb, $T = 27°F = 270°K$, $\rho = 1.30 \times 10^{-3}$ gm cm^{-3}, the pressure gradient force per unit mass, from Equation 1, is

$$|F_x| = \frac{1}{1.30 \times 10^{-3} \text{ gm cm}^{-3}} \cdot \frac{21 \times 10^3 \text{ dynes cm}^{-2}}{700 \times 10^5 \text{ cm}}$$

$$= 0.23 \text{ dynes gm}^{-1}$$

This force acts perpendicular to the isobars toward low pressure, that is, toward the west. A fundamental law of physics, Newton's second law, states that

$$F = ma \qquad (2)$$

The force F is in dynes when the acceleration a is in cm sec^{-2}. In the present example

$$a = 0.23 \text{ cm sec}^{-2}$$

If the force acts on the parcel for one day, or 8.64×10^4 sec, the resultant velocity is 200 m sec^{-1} (450 miles per hour).

The pressure gradient on the map of 8 January 1969 is not excessive and quite typical of the ones found on weather maps. Yet, according to the calculations, the pressure gradient force acting alone on a parcel of air for a day will develop a velocity never measured on the earth's surface. Obviously, other forces must be acting to balance the effect of the pressure gradient force.

7.5 *Rotation of the Earth, the Coriolis Force*

The facts that the winds do not reach the speeds calculated in the preceding section, and that the pressure gradients maintain themselves from day to day give evidence that forces other than the one of the pressure gradient are acting. The most important arises because of the rotation of the earth.

To understand better the effects of rotation, consider a merry-go-round rotating counterclockwise with four people on or near it (Figure 7.7). A is at the center, C off the rim, B on the wheel and between A and C, and D is ahead of B but on the same circle. Assume that A throws a ball toward B. C observes the ball travel in a straight line toward him. To both A and B, the ball does not travel in a straight line but deviates to the right of the direction of motion. This happens because to both A and B the direction changes with the rotation of the wheel. If B throws the ball toward A, the motion of the wheel still influences the ball after release and, once again, it will veer to the right of the path along which it was aimed. To C, the ball again takes a straight path but not toward A.

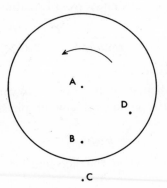

FIGURE 7.7 Rotation on a merry-go-round.

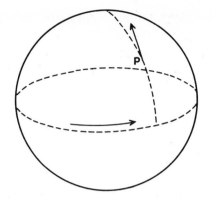

FIGURE 7.8 Motion of a particle northward on a rotating earth.

If *B* throws toward *D*, again the spin of the wheel makes the ball take a direction between the chord *BD* and the tangent at *B*. Because *D* also moves during the time of flight, the ball will appear to him to travel farther from a straight line, although again *C* sees the ball traveling straight. To *A*, *B*, and *D*, the deviations from a straight line could be explained by assuming that some other force pulled the ball to the right of its direction of motion.

A consideration of motion on the earth shows that the rotation of the earth has somewhat the same effect as the rotation of the merry-go-round. The exact derivation is beyond the scope of this book but Appendix 3 presents an approximate derivation of the resultant formula. If a body at *P* (Figure 7.8) on the earth is moving directly northward, it passes into a region where the speed of rotation is slower. Moving more rapidly eastward than the earth beneath it, the body arrives at a point to the east of the point *Q* directly north of *P*.

A body moving eastward rotates around the axis of the earth at a speed equal to the speed of rotation of the earth underneath plus its own speed relative to the earth. The centrifugal force acting on the body, a force proportional to the square of the speed, is greater than that on the earth beneath because of the body's motion. Part of this additional centrifugal force acts perpendicular to the earth's surface and reduces the gravitational force. This has little significance. The other part of the force acts along the surface of the earth and pulls the body southward. When the body is moving directly westward, the centrifugal force is less than that of the body stationary on the earth. The deficiency results in a net force moving the body northward. In both cases, the body deviates to the right of its path.

Careful analysis (see Appendix 3) shows that the apparent force on a moving unit mass is

$$F = 2\Omega v \sin \phi \tag{3}$$

where Ω is the angular rotation of the earth, 7.29×10^{-5} radians sec^{-1} (see Section 1.3), ϕ is the latitude, and v the speed. This force always acts perpendicular to the direction of motion, to the right in the Northern Hemisphere and to the left in the Southern Hemisphere.

The force of Equation 3 is named the Coriolis force after the French physicist, G. G. Coriolis, who discussed, in 1844, the apparent force arising from the rotation of the coordinate system. It will be denoted here as C. To save repetition, our discussion in this text will assume that the parcel is in the Northern Hemisphere unless otherwise stated. Conclusions can be readily transferred to the Southern Hemisphere if we remain aware of the change in the direction of action of the force.

The Coriolis force acts on all moving particles except the ones on the equator, but is so small that it becomes apparent only if the velocity is very large or if there is no force, such as friction, to oppose the motion. Even then, the effect is small unless it has opportunity to act for a finite period of time. Allowance must be made for its effect on artillery shells, although on rifle bullets the effect is negligible. Water in a river tends to wear down the right-hand bank more rapidly than the left-hand bank. In a bath tub, the small effect of the Coriolis force is sufficient, if no other forces are present, to cause the flow to deviate slightly from the straight line to the outlet, and the water acquires a circular motion counterclockwise as it approaches the hole. Except in special situations, the effect of the Coriolis force can be neglected. But the flow in the atmosphere is one of the special cases.

7.6 Geostrophic Winds

At the equator, the latitude is zero, and so the Coriolis force is zero. Any horizontal pressure gradient will cause particles of air to move toward the lowest pressure. With no force except friction to oppose the motion, the particles will leave the region of high pressure, reducing the excess, and move into the region of low pressure, reducing the pressure deficit. Eventually the flow will reduce the pressure gradient to zero. Centers of high or low pressure cannot persist at the equator.

Figure 7.9 illustrates the situation when a pressure gradient exists away from the equatorial region. A stationary parcel A, acted on by the pressure gradient force a will begin to move perpendicular to the isobars. This situation was discussed in the example of Section 7.4. Im-

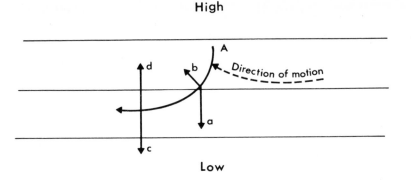

FIGURE 7.9 Motion of a particle, initially stationary, in a region of a horizontal pressure gradient; *a* and *c* are pressure gradient forces, *b* and *d* Coriolis forces.

mediately, the Coriolis force *b* begins to act perpendicular to the motion, causing the parcel to deviate toward the right of its path. With the change in direction, only part of the pressure gradient force is acting along the direction of motion to increase the speed. Gradually the direction changes, and the speed increases until the parcel is moving at a steady velocity parallel to the isobars, the pressure gradient force *c* acting to the left equaling the Coriolis force *d* to the right. There is no net force, and hence no further change of direction or speed. For this uniform motion, the balance of forces is given by

$$2\Omega v \sin \phi = \left| \frac{1}{\rho} \frac{\Delta p}{\Delta x} \right| \tag{4}$$

where now x is measured perpendicular to the isobars and *the direction of the wind is such that, standing with one's back to the wind, low pressure is to the left*. This rule goes by the name of Buys-Ballot's law from the name of the meteorologist who first proposed it in 1857.

In Equation 4, there are four variables, that is, v, ϕ, ρ, and $\Delta p/\Delta x$. In general, on the surface of the earth, variations in ρ are minor compared with the other variables. If a mean value is taken, the equation states that for any given latitude, there is a unique velocity that gives a balance between any pressure gradient force and the resulting Coriolis force.

In the example of Section 7.4, the pressure gradient force was shown to be 0.23 dynes gm^{-1}. At latitude 40°,

$$2\Omega \sin \phi = 9.37 \times 10^{-5} \sec^{-1}$$

For this latitude, the equation is satisfied if

$$v = 25 \times 10^2 \text{ cm sec}^{-1}$$
$$= 25 \text{ m sec}^{-1} \ (55 \text{ mi hr}^{-1})$$

The value of the wind velocity that satisfies Equation 4 is called the *geostrophic wind*. The quantity Δp is usually taken as the pressure interval between isobars, and Δx then becomes the distance between them. The geostrophic wind at a specified latitude is, then, inversely proportional to the isobar spacing. Near the earth, the variations in density may be ignored but this is not true in the free atmosphere.

Careful observations in the free atmosphere show that the wind deviates at times, but only slightly, from the geostrophic wind. In general we can assume that the wind velocity is given by the *geostrophic wind equation*, that is by Equation 4.

In Equation 4, if the direction along which the pressure change is measured is perpendicular to the isobars, the computed wind is the total wind. If another direction is taken, the computed wind is the wind component perpendicular to that direction. If the gradient is measured along two perpendicular lines, the two computed wind components may be combined by the method of the parallelogram of forces to obtain the resultant wind direction and speed.

The relationship given by the above equation is of vast importance in meteorology. The winds of the atmosphere are the means by which warm air or cold air, dry air or damp air, moves from place to place. They provide the method (discussed in Chapter 1) by which the atmosphere distributes the solar energy over the world. But wind velocities, because of their variability, are difficult to measure exactly and to combine into a consistent picture. Yet, because of the validity of Equation 4, it is possible to use the pressure of the atmosphere, which can be measured with extreme precision, to determine a close approximation to the mean winds.

Our weather maps, both current and forecast, are maps of the pressure distribution over the region. They are interpreted with the help of the geostrophic wind equation to give the wind flow, since the isobars can generally be considered the lines of flow. Other laws are then invoked to estimate the values of the other weather elements.

When the winds are geostrophic and, therefore, blowing along the isobars, there is no convergence or divergence.[1] Cyclones and anticy-

---

[1] As with many absolute statements in meteorology, this is not quite true. Some convergence or divergence results from variations in latitude. The effect is small but plays a part in the movements of cyclones and anticyclones over the earth.

clones can persist for days in middle and high latitudes. They are horizontal eddies—some 1000 km across and some 1 km diameter or smaller. Each one plays a part in carrying the solar energy from the equatorial regions poleward or into the upper levels of the atmosphere. The Coriolis force, which helps to control the winds, plays an important part in determining the weather processes on the earth.

Notice that Buys-Ballot's law is valid for cyclones when the wind blows counterclockwise in the eddy, and for anticyclones when it blows clockwise. In the Southern Hemisphere, the winds blow in the opposite directions.

7.7 Gradient Winds

In the circular flow such as is found in whirlwinds, tornadoes, and hurricanes, a force other than the two discussed in the last section comes into effect. Consider the flow around a center of low pressure (Figure 7.10). The pressure gradient force F on a particle P acts inward along the radius. The Coriolis force C, acting outward along the extended radius, requires that the motion of P be counterclockwise. Because the particle has a circular motion, there is a centrifugal force Ce which also acts outward. The value of this force is v^2/r where r is the radius. For a balance of forces

$$\left| \frac{1}{\rho} \frac{\Delta p}{\Delta x} \right| = 2\Omega v \sin \phi + \frac{v^2}{r} \tag{5}$$

This *gradient wind equation* can be solved for v, the velocity.

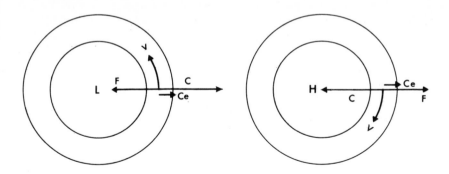

FIGURE 7.10 Forces on a moving particle in the vicinity of a low-pressure center; F, pressure gradient force; C, Coriolis force; Ce, centrifugal force.

FIGURE 7.11 Forces on a moving particle in the vicinity of a center of high pressure: F, pressure gradient force; C, Coriolis force; Ce, centrifugal force.

By the algebra of quadratic equations, this equation has two solutions, one giving v positive and one negative. The negative velocity would represent clockwise motion. This would mean that the centrifugal force balanced the Coriolis force and the pressure gradient force combined. Whirlwinds and tornadoes have been observed in which the wind blew clockwise, but this is unusual, at least for larger storms.

Without solving the gradient wind equation, one can observe that when r is large the last term is small. The velocity necessary to satisfy the equation is then very nearly, but slightly less than, the geostrophic wind velocity. This means that in cyclonic flow, the wind is less than the value derived from the geostrophic wind equation. When r is small, the reduction below the geostrophic wind can be considerable.

When the wind is blowing around a center of high pressure (Figure 7.11), the balance of forces is changed. The pressure gradient force F now acts outward, and the Coriolis force C acts toward the center. Therefore

$$2\Omega v \sin \phi = \left| \frac{1}{\rho} \frac{\Delta p}{\Delta x} \right| + \frac{v^2}{r} \tag{6}$$

To satisfy this equation, the value v must be slightly higher than the geostrophic wind value. Around anticyclones the radius of curvature is usually large and the correction to the geostrophic wind is seldom more than 10 per cent. This is in contrast to the situations around low-pressure areas. In a hurricane the major rotation is restricted to an area considerably less than 200 km in radius. The wind, high though it is, is reduced considerably from the geostrophic value determined from the pressure gradient.

In the foregoing discussion in this section and also in Section 7.5, we have treated the centrifugal force as if it were a real force on the moving particle. In reality, there is no such force on the particle. Consider the motion of a body rotating at the end of a string. If the string breaks, the body does not move outward, which it would do if the force were real, but along the tangent to the circle.

When a body moves under the influence of a net force which is proportional to the square of the speed and perpendicular to its direction of motion, it rotates in a circle. One may then consider that the body is rotating under a balanced system with a force acting along the extended radius equaling V^2/r. It is with this point of view that the discussions concerning the rotation of the earth and the rotation about a pressure center have introduced the centrifugal force. Some theoretical physicists have pointed out the error in such a treatment, but it seems to make the causes of the changes in the velocity around a pressure center and other observable results easier to understand.

Under most flow conditions, the difference between the gradient wind and the geostrophic wind determined by the spacing of the isobars is small, and the wind still blows parallel to the isobars. Therefore, the remarks about the relationship between winds and the pressure distribution at the end of Section 7.6 are still valid. The isobars still give approximately the direction of flow of the air even when they are curved.

Gradient winds are found when the flow is no longer straight, but when there is a tendency for a rotary motion. The presence of such a motion has significance in meteorology because, according to a law of fluid motion, rotary motion has a tendency to maintain itself. Friction will gradually reduce the motion and the vortex, but this acts slowly over uniform surfaces and other processes can counteract the effect.

An example of vortices can be observed downstream from a rock or pier in a river. An eddy forms just below the rock and then maintains itself for some distance as it leaves its place of origin and moves in the general flow. In a similar manner, a vortex forms around the corner of a building and floats down wind. Occasionally, we see them in the form of whirlwinds but many eddies, some large and some small, are never seen. Low-pressure centers and anticyclones are examples of very large eddies within the atmosphere. Meteorologists measure this rotary motion by a quantity called *vorticity*. Cyclonic vorticity occurs with counterclockwise motion and has, by convention, a positive value; anticyclonic vorticity is negative.

The conservation of rotary motion in a fluid acts on the vortices that form our highs and lows. Just as with eddies in a stream, they move with the motion of the air aloft, carrying their weather phenomena along with them. Usually, with cyclonic vorticity, there is divergence aloft, which tends to reduce the central pressure and so counteract some of the effects of friction (see Section 7.8). Convergence aloft above a region of anticyclonic vorticity maintains the center of high pressure and permits the anticyclone to move as a unit downwind with only a slow change in its characteristics.

7.8 *The Effect of Friction on Winds*

Another force that is found in the atmosphere, helping determine the wind speed and direction, is the force of friction, or the retarding effect of trees, buildings, and other irregularities in the topography. They tend to reduce the speed of the wind, but this reduction has

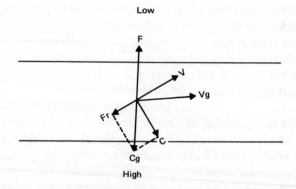

FIGURE 7.12 Effect of friction on the geostrophic wind: F, pressure gradient force; C, Coriolis force; Fr, force of friction.

other consequences. Consider Figure 7.12. The force F, representing the pressure gradient force, can be balanced by $C_G = 2 \, \Omega v_G \sin \phi$, the Coriolis force if the wind is blowing parallel to the isobars at the geostropic wind velocity, v_G. A force of friction Fr reduces the speed to v. This then reduces the Coriolis force to $C = 2 \, \Omega v \sin \phi$. Lacking a balance, the pressure gradient force begins to pull the particle across the isobars toward lower pressure. A balance will again be reached when the wind v is blowing across the isobars at an angle α. The force Fr, acting directly opposite to the wind, and C, acting normal to the wind, add together by the law of the parallelogram of forces to be equal to C_G.

An appreciation of the effect of friction can be obtained by considering the movement of a car on a well banked curved path. Here, the centrifugal force acting outward is able to counteract the force of gravity, which would carry the car inward down the slope of the track. If suddenly the car slows down, the centrifugal force can no longer balance the force tending to carry the car downhill and, hence, the car can no longer maintain its altitude.

Two results can be observed from the discussion. When friction is small, the flow differs very little from the geostrophic; when large, the relationship between isobars and winds can be slight. Observations confirm these conclusions. Over snow surfaces and calm waters, winds are close to the geostrophic both in direction and speed and give reliable clues to the direction and spacing of the isobars. In mountainous regions, the winds are light and intersect the isobars at large angles. An examination of an analyzed weather map confirms these conclusions. Over the oceans, the plotted winds are at small angles to the isobars. Over land areas with few major topographic features, such as the Mississippi Valley, the winds still blow somewhat along the isobar but at a greater angle. In the mountainous terrain of the western United States and elsewhere, the surface wind directions have little relationship to the sea-level pressure distribution.

Friction causes a movement of air across the isobars toward low pressure. Unless there is some compensating effect aloft, this flow will add to the mass at the center of the low, and so reduce the pressure gradient. This result can at times be noted in a hurricane. At sea, the inflow toward the center must be balanced by a divergence at high levels if the hurricane is to maintain itself. When the storm crosses a coastline, the surface below has become suddenly rougher and drier. The additional inflow toward the center is not balanced aloft, and air begins to fill the surface low. With the reduction of pressure comes a reduction in wind, and the destructive force of these storms rapidly decreases. The decrease in the supply of moisture is another reason for the rapid drop in the destructive force.

7.9 Upper-level Charts

The effect of the earth's surface in hindering air movement decreases with altitude, so that above 0.5 km over relatively smooth ground it may be neglected. The actual depth varies. Under stable conditions over smooth ground, the wind increases rapidly with height, and the *friction* or boundary *layer* is shallow. It tends to be deep over rough terrain or under unstable conditions. The winds at the top of the friction layer are usually close to the gradient or geostrophic velocity, and they can be determined with sufficient accuracy by the equations given in Sections 7.6 and 7.7.

Initially, the purpose of observing winds in the free atmosphere by balloons, or computing them from pressure maps for 5000 ft, etc., was to provide data to artillery officers and later to aircraft navigators. As meteorologists became more familiar with winds aloft during the early 1940's, they discovered that the flow at upper levels had significance in projecting weather patterns. For instance, they discovered that highs and lows tended to move with the velocity of the winds at 10,000 ft. Gradually, the interest in these levels has increased, so that now maps showing the wind flow to 10 km are standard in all weather offices and, in some forecast offices, maps to 16 km are drawn.

Maps for upper levels differ from the ones for sea level in the units used. Figure 7.13, in conjunction with Figure 3.8, will help to explain the difference. Figure 3.8 gives the isobars for 1200 h GMT 8 January

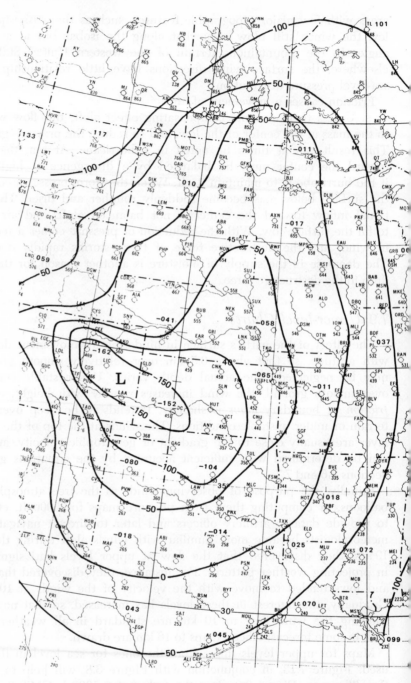

FIGURE 7.13 Height of the 1000-mb surface at 1200 h GMT 8

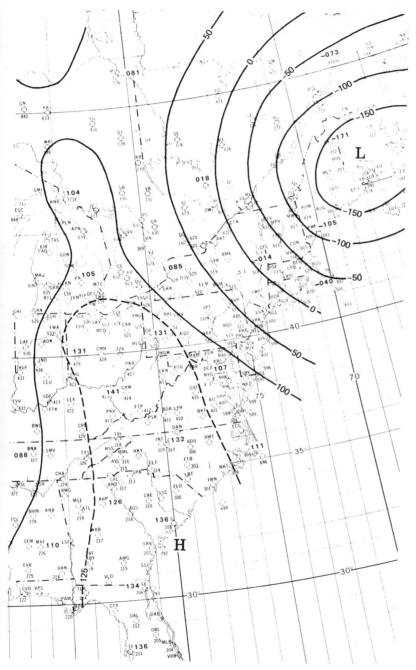

January 1969. (Compare with Figures 3.8 and 10.2.)

1969 for part of eastern and central North America. The sea-level pressures varied from 1018 mb in central Saskatchewan to 985 mb in eastern Colorado and 977 in New Brunswick. The 1000-mb curve enclosed much of the Great Plains west of the Mississippi River with a second curve through eastern New York State. Where the sea-level pressure was above 1000 mb, the height of the level at which the pressure was 1000 mb could be computed. Maximum values were approximately 130 m, found in Saskatchewan and from Lake Erie south to Florida.

On the average, the height in meters at which the pressure is 1000 mb is 8.5 times the excess of pressure above 1000. A correction is related to temperature and becomes significant in regions of high pressure. From this consideration, we can visualize an invisible surface along which the pressure was 1000 mb. Figure 7.13 shows the zero-height, 50-m, and 100-m contours of this surface. The zero-height is, of course, the 1000-mb curve for the sea-level pressure map.

If the station has a sea-level pressure below 1000 mb, it is necessary to say that the height of the surface is negative. Thus at Denver, Colorado the height was −162 m, and at Caribou, Maine, −171 m. Figure 7.13 shows the −50-m, −100-m, and −150-m contours. The total picture of the 1000-mb surface (Figure 7.13) differs only slightly from the sea-level map (Figure 3.8).

This discussion shows that it is possible to have weather maps giving the contours of the 1000-mb surface that differ in appearance only slightly from maps of sea-level pressures. Where pressure is high, the surface is high; where low, the surface is low. Maps of contours of a constant-pressure surface are drawn at times for 1000 mb and regularly for 850 mb, 700 mb, 500 mb, 300 mb, and sometimes for other values. These maps are approximately 1.4 km, 3 km, 5.5 km, and 9 km, respectively, above sea level.

The use of constant-pressure maps simplifies the geostrophic wind equation. According to Equation 4, Section 7.6,

$$v = \frac{1}{2\Omega \sin \phi \rho} \left| \frac{\Delta p}{\Delta x} \right| \tag{7}$$

Consider Figure 7.14, where A and B are two points on a constant-pressure surface, AC is horizontal, and BC is vertical. Then the wind velocity normal to AC is given by

$$v = \frac{1}{2\Omega \sin \phi \rho} \frac{p_C - p_A}{AC}$$

According to the hydrostatic equation

$$p_C - p_B = g\rho BC$$

Because $p_A = p_B$, then

$$v = \frac{g}{2\Omega \sin \phi} \frac{BC}{AC} \tag{8}$$

The fraction BC/AC is the slope of the constant-pressure surface, and may be obtained from a contour map of the surface even as $\Delta p/\Delta x$ may be obtained from a constant-level map. By using the normal coordinate system, the fraction is written $\Delta z/\Delta x$.

The advantage of Equation 8 over Equation 4 for giving the geostrophic wind is in the elimination of density in the computation. Thus the same scale may be used to convert gradient on a constant-pressure map to the geostrophic wind, whether the map is for the 850-mb surface or for the 50-mb surface. For any latitude, the spacing of the contour lines gives the wind speed directly.

It is possible to compare the winds on a constant-pressure surface with the motion of a car around a banked curve, although the forces are not the same. With the car, the speed of motion for a balance must increase with the slope of the road. So, too, must the speed of the wind increase to give a balance as the slope of the constant-pressure surface increases. Under conditions of balance, both wind and car maintain the same altitude. This means that, in the atmosphere, the winds blow parallel to the contours.

7.10 Thermal Winds

The upper-level charts, when they were first drawn, showed clearly what watchers of clouds had known for a long time, that the winds aloft are not always in the same direction as the ones at the surface of the earth. This can be observed by reference to Figure 7.15. According to the preceding section, the wind speed and direction is determined by the slope BC/AC of the surface (see Figure 7.14). If (Figure 7.15) AB is the 700-mb surface, the slope of AB determines the horizontal wind perpendicular to AB. If the air over B is warmer than that over A, by the hydrostatic equation Δz between the 700- and 500-mb surfaces

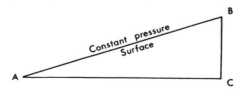

FIGURE 7.14 Computation of the geostrophic wind on a constant-level surface.

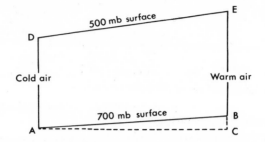

FIGURE 7.15 Effect of a temperature gradient on the slope of a constant pressure surface.

is greater over B than over A. If DE is this 500-mb surface, then $AD < BE$. Therefore the slope of DE will be greater than the slope of AB and the wind perpendicular to DE will be greater than that perpendicular to AB.

This illustrates the effect of a temperature difference on the change of wind with height. The amount of the change, usually represented by the vectorial difference between the winds at two levels, is called the *thermal wind*. In Figure 7.16, if V_1 represents the velocity of the wind at level z_1 and V_2 the velocity at $z_2 > z_1$, then the thermal wind is given, in speed and direction, by the vector V_T. This thermal wind is closely related to the mean temperature of the layer z_1z_2, being parallel to the isotherms for the layer.

Figure 7.16 suggests that the thermal wind is derived from the actual winds at two different levels. In practice the upper-level wind is sometimes deduced from the lower-level wind and the thermal wind. Thus, in Figure 7.16, knowing V_1 and V_T, one determines V_2. It can be shown that V_T is parallel to the isotherms of the mean temperature, and proportional to the horizontal temperature gradient with cold air to the left looking downwind, in much the same manner as the geostrophic wind is parallel to the isobars and proportional to the pressure gradient. Also, the thermal wind is proportional to the distance between the levels.

On the basis of these facts, we reach some general conclusions on upper winds:

1. As one goes farther aloft the winds tend to parallel the isotherms more closely. Because in the middle and upper troposphere the equator is warm and the poles cold, the winds at these levels tend to be westerly.

2. Above a center of low pressure which is cold relative to the surroundings, the two winds augment each other and the circulation increases aloft.

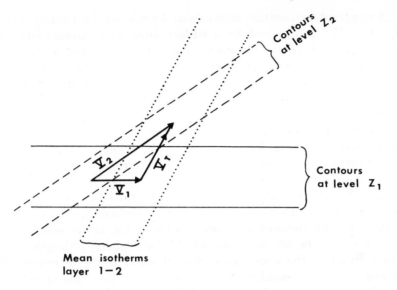

FIGURE 7.16 Relationship among the geostrophic winds at two levels of the atmosphere. V_1 and V_2, and the thermal wind, V_T, for the intervening layer.

3. As one rises in the atmosphere, the center of a low tends to shift toward the area of coldest temperature.

4. Above a warm anticyclone, the anticyclonic winds increase with height.

5. The anticyclonic flow aloft tends to die down above a cold anticyclone.

7.11 The Jet Stream

One interesting and significant development arising from the results of the last section is the formation of the jet stream. The effect of the temperature pattern gradually outweighs the surface pressure pattern in determining the winds above 5 km. In one type of situation, discussed in Chapter 10 and elsewhere, the temperature gradient is large (see Figure 2.9). This occurs when cold air and warm air lie side by side, forming a *front*. The isotherms are parallel to the front, and with strong fronts the pattern persists to the tropopause, although with a decreasing horizontal gradient. Under these circumstances, the effect of the temperature gradient is cumulative so that the upper wind becomes parallel to the front and increases in speed with height.

A temperature contrast in northern Florida on 20 January 1958 (see Figure 2.15) was caused by a surface front (see Section 10.4) which sloped upward toward the north. The thermal contrast associated with the front produced a jet stream over South Carolina. This is shown in Figure 7.17 which gives the isolines of speeds of the westerly components of the winds from 20°N to 50°N at 80°W longitude. Notice that the wind speed reached a maximum of more than 60 m sec^{-1} (130 mi hr^{-1}) in a core found under the tropopause and at an altitude of approximately 11 km. The decrease in the wind above the tropopause is normal. In the stratosphere, the temperature gradient usually reverses itself, with relatively warm stratospheric temperatures above the cold air (see Figures 2.15 and 2.16). Thus the thermal wind acts in opposition to the wind of the troposphere and, thereby, reduces it.

A typical jet stream has a core 200 km or less across and 3 km deep. This core can be followed around the hemisphere, although the maximum speed in the core varies. Sometimes it splits, to reunite farther to the east. The boundary of the core, where the wind speed changes rapidly with distance, is a region of high turbulence which can produce vibrations in an aircraft flying there. As mentioned before, it is associated with fronts. It also appears to influence the development of tornadoes and other severe convective storms Possibly, vertical currents in the vicinity of the jet stream augment vertical currents produced by instability to cause storms to increase in intensity.

7.12 Land and Sea Breezes

An example of a local wind initiated by pressure differences is found in the land and sea breezes of coastal areas. Assume that in the early morning along a seacoast the pressures are uniform, so that the pressure surfaces (Figure 7.18) are level. The land warms more rapidly than the sea, and so above it the constant-pressure surfaces move upward. The slope of the surfaces now initiates a wind seaward above 500 m. The change in distribution of mass arising from this movement raises the pressure surfaces over the ocean below 200 m and lowers them over the land. The sloping pressure surfaces induce a return current near the surface which brings cool air across the coast, producing the afternoon sea breeze. Mean pressure observations along the British Columbia coast confirm these pressure changes. At Cape St. James, Queen Charlotte Islands (52°N 131°W), the mean July pressure for 21 years for 16 h PST was 1.0 mb higher than the mean pressure at 04 h. During the same 12 hours, that is, between 04 and 16 h, the

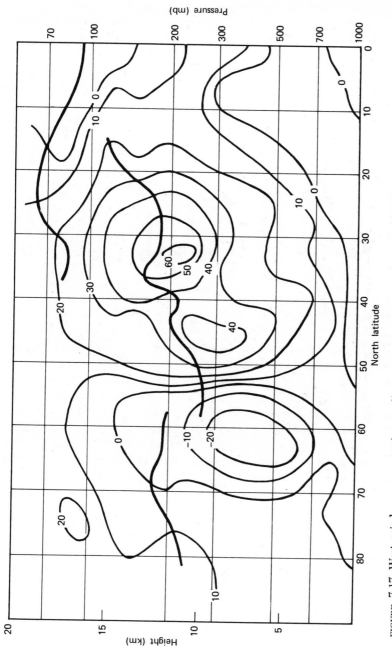

FIGURE 7.17 West wind components (m sec^{-1}) in a cross section of the atmosphere along 80°W at 1200 h GMT 20 January 1958, showing a jet stream. (Compare with Figure 2.15) Heavy line gives the tropopause. (From U.S. Department of Commerce, Weather Bureau, 1963: *Daily Aerological Cross Sections for the IGY Period.* Washington, D.C.

163

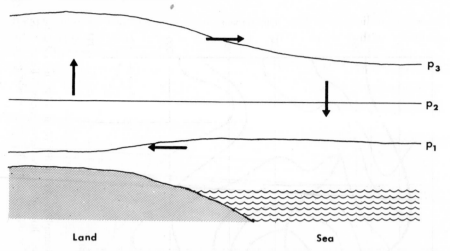

Land   Sea

FIGURE 7.18 Pressure surfaces and winds across the boundary between a cold ocean and warm land.

mean pressure at Lytton, in the Fraser River Valley (50°N 112°W), dropped 3.7 mb.

Initially, the Coriolis force does not act, and the winds tend to blow normal to the coastline. As the afternoon passes and the wind increases, the effect of the Coriolis force becomes sufficient to cause the wind to blow at a smaller angle to the coast. The upper-level current blowing seaward is distributed over a thick layer and is not easily detected. The surface flow is chiefly in the lowest kilometer, and its arrival is marked by an abrupt wind shift and temperature drop.

The temperature contrast reverses itself during the evening, and at night the land is normally colder than the ocean. This produces a reversal of the surface wind, but the land breeze of the night is not so pronounced as is the sea breeze of the afternoon.

The strength and depth of penetration of the sea breeze depend on many factors. Of them, the temperature contrast across the coastline is the most important. In the tropics, the sea breeze is a diurnal phenomenon and can penetrate 20 km inland. In temperate latitudes, the effect can occasionally be felt 100 km inland, but usually it is restricted to 30 km or less. If there is an initial wind flow before the heating of the day begins, the effect is to augment or counteract this initial flow.

Figure 7.19 shows an example of the sea-breeze winds. The cross section goes from a ship off Block Island, Rhode Island, inland on a line perpendicular to the coast. The lines are isolines of wind speed in knots with positive values indicating a wind blowing toward the land. The observations were taken on 2 August 1958, a day of sunny skies and initially light winds. The winds near the surface increased from a maximum of 8 kts at 0900 h EST to 28 kts at 1700 h EST.

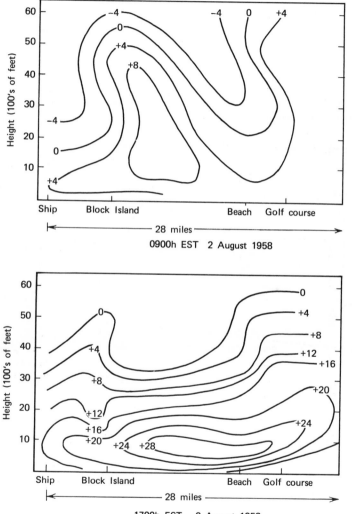

FIGURE 7.19 Development of a wind off the coast of Rhode Island. Wind velocities (knots) normal to the coast with positive values blowing from sea to land. (After Fisher, published by permission of Mrs. Edwin L. Fisher. From Fisher, Edwin L., 1960: An observational study of the sea breeze. *J. Meteor.*, 17, 650.)

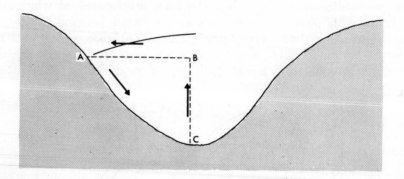

FIGURE 7.20 Pressure surfaces and winds in a cross section of a valley during a night of radiational cooling.

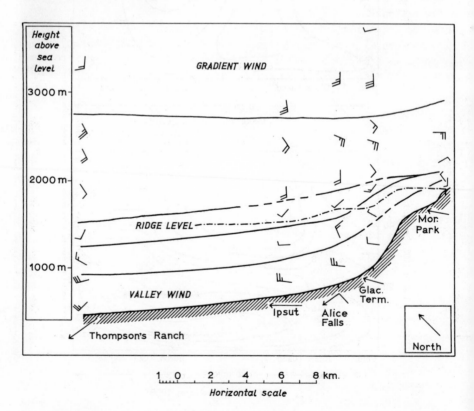

FIGURE 7.21 Mountain and valley winds. Isopleths of valley-parallel wind components (m min⁻¹) in upper Carbon River Valley, Washington, 1400 h PST 17 August 1958. (After Buettner and Thyer. Published by permission of K. J. K. Buettner. From Buettner, K. J. K., and N. Thyer, 1959: *On valley and mountain winds.* University of Washington, Department of Meteorology and Climatology, p. 15.)

FIGURE 7.22 Oil fog plumes from three generators slowly flowing downhill in the early evening. Notice the wind shear on the plume to the right. Geostrophic wind was from right to left but a valley rotor caused shear from left to right on the plume. (Courtesy R. M. Holmes. From Holmes, R. M., 1969: *The Climate of the Cypress Hills*. Canada, Department of Energy, Mines and Resources, Calgary, Alberta, 42 pp.)

The return flow above 4000 ft shows in the diagram only as very light but, as shown in Figure 7.19*a*, the wind aloft in the early morning was light onshore. Thus the return flow was evident only in the decrease of the normal flow at upper levels. The observations showed that the Coriolis force influenced the winds slightly during the day, causing them to blow more nearly parallel to the coast than they otherwise would have done.

importance is minor. Yet the principles involved (the differential heating A sea breeze is a local phenomenon, and it would appear that its of water and land and the resultant pressure changes) have widespread significance. For example, in the annual rhythm of temperature, a body of water like Lake Superior is colder than the surrounding land during the summer and warmer during the winter. Studies of the winds around the lake have shown that summer winds tend to be onshore, and winter winds are offshore. Therefore, they are similar to the day and night winds, respectively, of the land-and-sea-breeze cycle.

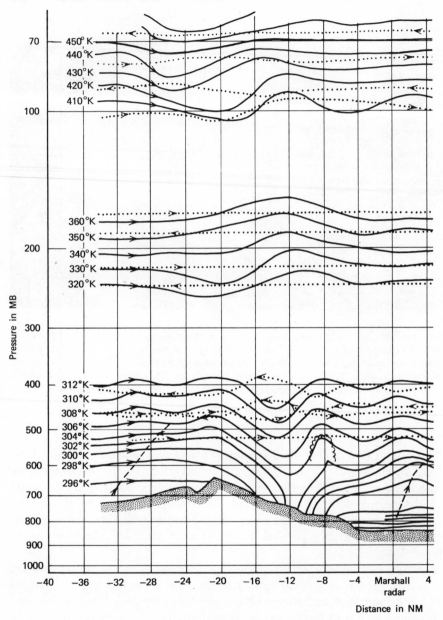

FIGURE 7.23 Flow patterns over and to the lee of the mountain near Boulder, potential analysis for 15 February 1968. The dotted and dashed lines represent From Kuettner, J. P., and D. K. Lilly, 1968: *Lee waves in the Colorado Rockies.*

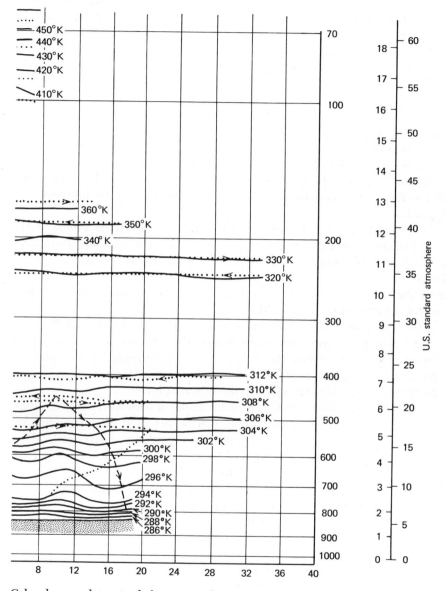

Colorado, as determined by an analysis of potential temperature. Preliminary aircraft and balloon flight trajectories. (Published by permission of J. P. Kuettner Weatherwise, 21, 196–197.

7.13 *Mountain and Valley Winds*

A second example of a local wind system initiated by pressure differences is found in many mountain valleys. They are called mountain and valley winds.

Consider the constant-pressure surfaces in Figure 7.20. Lacking any overall pressure differences, the surfaces in the free atmosphere will be level. Radiational heat exchange at A (at the surface) will cause greater temperature changes than at B, in the free atmosphere but at the same level. When A is cooling, the pressure surfaces will dip downward near A, initiating a flow of air from B to A, downward along the valley side from A to C, and upward from C to B. When incoming radiation is greater than outgoing, the result is a reversal of the flow. Of the flows, the one most readily noticed is that along the hill between A and C.

An example of the daytime valley winds is shown in Figure 7.21. This presents a picture of the winds observed at a number of stations in Carbon River Valley, Washington at 1400 h PST 17 August 1958. There was a strong up-valley wind in the lowest 700 m, above which layer the winds became irregular, finally becoming south under the influence of the overall pressure field.

An example of a nighttime valley wind is shown in Figure 7.22. This picture was taken from an aircraft during the early evening in southern Alberta. The flows down the valleys are made visible by the plumes from some oil flares. Part of a third plume rose high enough to be torn apart by the different winds in the different levels.

Mountain and valley winds are found under a wide variety of circumstances. At times, a slight downwind can be detected after sundown in the lowest three feet above a slope. One gardener discovered, after building a stone wall at the bottom of his garden, that he had dammed up this flow and increased the risk of frost to his crops. Light down-valley winds have been observed to move out over the plain, and then to be caught in a vertical flow as they moved over a warm river. On the other end of the scale, the glaciers of Greenland and Antarctica are the causes of strong *katabatic* winds down the fiords at the boundaries of the ice caps. Up-valley winds are not so well documented, but the banner clouds above many mountain peaks are the result of slope winds carrying warm air up the sides of the mountains.

When a wind blows perpendicular to a mountain range, the air must rise on the windward side. To the lee of the ridge the air subsides, and this motion sets up a standing wave that persists with decreasing amplitude downwind from the ridge (see Figure 7.23). Under certain circumstances, the air becomes saturated at the top of its path and, hence, a cloud forms. Thus we see a number of parallel lines of clouds at the wave crests. Glider pilots have been able to investigate these standing waves. In the region of rising currents they are able not only to stay aloft but also to climb. According to their records, the vertical currents induced by the mountain range can persist to more than double the height of the mountain.

The downdraft in the lee of the ridge has been found to be, at times, extreme. Aircraft caught in this current have experienced extreme jolts in the turbulence associated with the downdraft. One pilot reported a drop of 800 ft in less than 5 seconds, followed by another drop shortly afterward. There is much yet to be learned about the effect of topography on the winds and turbulence in free air above and near them.

PROBLEMS AND EXERCISES

1. Compute the distance between isobars at 4-mb intervals at a latitude of 40° for a geostrophic wind speed of 10 m sec^{-1} and a density of 1.23×10^{-3} gm cm^{-3}. If the map scale is $1:10^{7}$, what would be the distance between isobars on the map? Starting with this information, make a scale showing the distance between isobars for 40, 30, 20, 15, 10 and 5 m sec^{-1}.

2. On the 500-mb constant-pressure surface, latitude 43.4°N, the distance between 80-m contours is 500 km. Compute the geostrophic wind.

3. On a weather map, notice the wind at a ship or an island station. Compare the speed and direction with the geostrophic wind as given by the isobars.

4. Compare the geostrophic winds for the same spacing of isobars and the same latitude if, in one instance, the temperature is 77°F and, in the second, −40°F.

5. Two points, A and B, are 200 km apart. At A, the sea-level pressure is 1000 mb and the temperature 15°C. At B, the corresponding values are 1015 mb and 17°C. Determine the slope of the 1000-mb surface between A and B, assuming that lapse rates at both locations are zero.

6. The height of the 700-mb surface is the same at A and B. What information does this give about the wind at that level? The mean temperatures for the layer between 700 and 500 mb are −5°C and −12°C, respectively. Determine the height difference of the 500-mb surface between A and B. If $AB = 200$ km, what is the speed of the wind at 500 mb normal to AB? (Assume that $\phi = 43.4°$.)

7. A projectile at 50°N moves south to 48°N at a speed of 333 m sec^{-1}. Determine the Coriolis force acting on unit mass. From the value, determine the distance west of the target that the projectile will land. Compute for the same projectile the distance it rotates eastward during the time of flight and the distance the target point rotates eastward. Using these values, determine again the distance the projectile lands west of the target.

8. Determine the difference in heights of the 700-mb and 500-mb surfaces if the mean temperature of the layer is −10°C. By what amount must the temperature increase in order that the height difference increase by 100 m?

CHAPTER 8

AIR MODIFICATION AND AIR MASSES

8.1 Modification of an Air Column by Advection

In Chapter 6, we observed that the vertical temperature and moisture distributions in an air column are factors in determining the weather at the base of that column. This vertical structure is continually changing because of various influences on it. Many of these processes have already been discussed, but it is well to examine their effects on any given air column.

For a specific location outside the tropical and equatorial areas of the earth, the greatest changes in temperature and moisture come through the horizontal movements of air masses, that is, by advection. For any parcel, advection in itself does not change the temperature or moisture, but it does carry the parcel into a different region where the processes of subsidence, radiation, and evaporation, (discussed in Sections 8.2 to 8.4) may cause a change. But advection can and does change the structure of the column if a *vertical wind shear* is present, that is, if the wind velocity changes as one ascends. The change that occurs within the column depends, of course, on the three-dimensional temperature, moisture, and wind structure. If the winds cause a drop in temperature aloft relative to the temperature near the earth, the

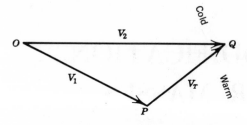

FIGURE 8.1 Advection in a wind shear.

stability of the column will decrease and the probability of convective storms will increase.

The advection of air is related to the thermal wind (see Section 7.10). If (Figure 8.1) V_1 is the geostrophic wind at z_1 and V_2 the wind at level z_2, PQ is the thermal wind that is parallel to the isotherms for the layer. According to the laws governing the thermal wind, cold air lies to the left of PQ looking downwind, and warm air to the right. Therefore, in the diagram, colder air is moving over O in the layer between z_1 and z_2. This will always be the situation when the wind backs (that is, turns counterclockwise) with height. When the wind veers (turns clockwise) with height, warm air is moving over the observing station. If the wind first backs with height and then veers, the air column at the station is becoming more stable. If the reverse is true, the instability is increasing with time.

8.2 Air Masses—Modification Through Subsidence

The modification that develops through advection only is usually local in nature. Meteorologists have been interested in processes by which a large quantity of air, of the order of 1000 km in diameter, acquires temperature and moisture values that are in equilibrium with the environment. When this equilibrium has been reached, the various processes tending to change the meteorological elements may still be active, but they tend to balance each other. The quantity of air is called an *air mass*. In an air mass changes of temperature and moisture in the horizontal are small. The vertical distribution is typical of the air mass. Because of its characteristics, an air mass is usually accompanied by a particular type of weather. The name attached to an air mass is often related to the district or zone in which it developed, that is, its *source region*. Thus, for example, we speak of tropical continental air or arctic air. A source region has a relatively uniform underlying surface, such as a large oceanic area or a continental plain.

Subsidence is a major process in the development of certain types of air masses. As described in Section 6.7, subsidence of a layer tends to produce stable, dry air. This is particularly true when the subsidence is accompanied by the horizontal divergence resulting from outflowing air, as in anticyclones. With subsidence, clouds evaporate and the sky is generally clear. In particular, subsidence in the subtropical anticyclones at approximately 25° to 40° latitude (see Figures 3.10, 3.11, and 4.16) tends to produce uniformly warm temperatures aloft. In the polar anticyclones of Siberia, northwestern North America, and Antarctica, the subsidence aloft tends to counteract the loss of heat by radiation in the lower and middle troposphere.

In contrast to subsidence, rising air tends to form clouds and rain. This situation is found in the vicinity of a low-pressure area. When the low lies outside the tropics, the different flows of air converging on the center have widely different characteristics. Meanwhile the time involved is not sufficiently long to eliminate these differences. Air masses do not develop within these systems.

The equatorial trough of low pressure (see Figures 3.10 and 3.11) differs from the extratropical lows. When the trough lies over the ocean, the converging flows of the northeast and southeast trade winds bring similar air of high temperature and moisture values into the trough. Here the air tends to be unstable or, at most, neutral because vertical currents caused by the convergence mix the air columns. Equatorial air is therefore moist, hot, and unstable.

8.3 *Radiation and Conduction*

Physically the two processes of radiation and conduction differ, but their effects on an air column are such that they may be considered together. Conduction heats or cools a very thin layer of air close to the earth's surface. The net effect on the air column of long-wave radiation from the earth, which is the most important radiation in determining the temperature, is also to heat or cool the lowest layer, but a thicker layer than the one affected by conduction. Other processes make the difference between the thickness of the layers of little significance.

If radiation is the only cause of temperature change, the temperature in the free atmosphere will cool at rates up to 2 deg C day^{-1}. The energy transferred from a warm earth to the lowest layers moves aloft by repeated radiation and absorption. This process is slow and, therefore, an unstable lapse rate will tend to develop. Before this stage is reached, vertical currents mix the air column, carrying warm air aloft and cool air down to the surface to become warmed.

The final temperature distribution depends on the surface conditions. Over desert areas, for example, Sahara or Arizona in summer, the air during the afternoon has a dry adiabatic lapse rate to 5 km or higher. Figure 8.2 gives the temperature and dew point curves for Tucson, Arizona for 5 July 1961, 1700 h MST, and shows the results of these processes. The curve 12 hours later shows how nocturnal cooling produced an inversion near the surface.

A snow surface absorbs little solar heat, and cools rapidly by black-body radiation as given in Equation 1, Section 5.2. Heat to warm the snow is transferred downward from the air column by both conduction and radiation. Because the air absorbs and radiates only in certain wavelengths, the heat emitted is less than that given by a black body at the same temperature. In order that a balance be reached between the upward flow from the snow and the downward flow from the air, the temperature of the snow surface must be less than that above. Computations show that the difference for a stable situation is about 20 deg C.

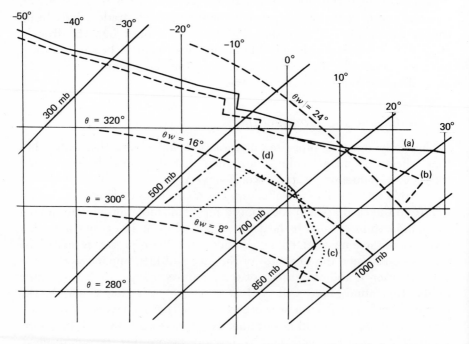

FIGURE 8.2 Summer air over Tucson, Arizona: temperature (*a*) and dew point (*c*) 1700 h MST 5 July 1961; temperature (*b*) and dew point (*d*) at 0500 h MST 6 July 1961.

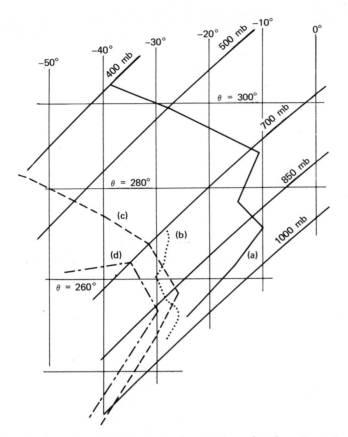

FIGURE 8.3 Typical temperature curves in polar continental and arctic air in winter: temperature (a) and dew point (b) at Bismarck, North Dakota; temperature (c) and dew point (d) at Eureka, Northwest Territories (80°N 86°W).

This means that normally the maximum increase of temperature in the inversion is 20 deg C. Of course, the warm layer is losing energy to layers of the atmosphere above it as well as to the earth, and will cool. Usually subsidence in an anticyclone compensates somewhat for the radiational cooling in the warm layer, and the inversion maintains itself. Figure 8.3 shows two typical ascents in anticyclones over snow-covered surfaces.

The historical development of the concept of air masses has resulted in a peculiar terminology. Initially, in the 1920's, air masses were described as being either polar or tropical. Further study brought a recognition of equatorial air and of other air masses that developed north of the "polar" air masses. These latter were called arctic, so that we have arctic air being colder than polar air, although geographically the pole is the northernmost point in the Arctic.

8.4 *Modification by Evaporation*

Evaporation from moist soil, plants, lakes, and the open ocean produces changes in the temperature and moisture of an air column. As discussed in Section 4.11, the rate of evaporation depends on the difference between the vapor pressure over the water and the vapor pressure in the air above it. The initial movement of moisture is into the lowest layers. An equilibrium would be reached if the air became saturated at the temperature of the surface water. But vertical currents, which develop because of the warm water (if the water temperature is above the air temperature), and mechanical turbulence as a result of the wind distribute the moisture through a thicker layer and permit evaporation to continue. The depth of the layer depends on the temperature distribution of the air column. Notice the moist layer near the surface in an air column over Midway Island in the Pacific subtropical high with the dry air aloft, as shown in curves *a* and *b* of Figure 8.4.

When the northeast trades carry tropical maritime air toward the equator, increasing surface temperatures with increased evaporation produce instability. At first, the vertical currents are topped by the stable air aloft, producing cumulus clouds, but gradually the moist layer becomes deeper. Curves *c* and *d* of Figure 8.4 show the situation near the equatorial trough where the modification of the tropical maritime air has produced equatorial air with its cumulonimbus clouds and showers. When tropical maritime air is carried northward into the Gulf of Mexico around the western end of the subtropical Atlantic or Bermuda high, modification through additional heating and evaporation produces changes, so that the air that reaches the Gulf Coast of the United States in summer is not greatly different from equatorial air.

Modification is extremely rapid in another common situation. When winter polar continental air or arctic air is swept over the open ocean, heat is transferred from the water through conduction and radiation to the lowest layers of the air. Also, evaporation from the relatively warm waters transfers much moisture into the dry air. This may occur off the east coast of Siberia or when air is swept off Alaska around a center of low pressure in the Gulf of Alaska. Arctic air also moves over the Atlantic from Labrador or Greenland. The surface heating eliminates the normal inversion, and the instability that develops may produce cumulonimbus clouds to 6 or 7 km. One reason for the high tops is

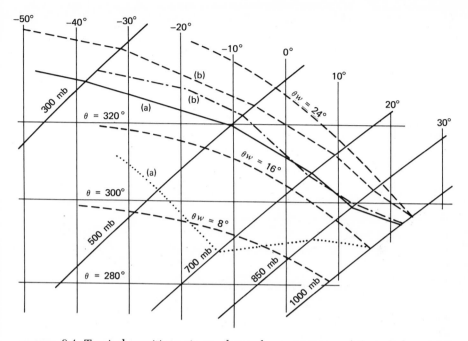

FIGURE 8.4 Tropical maritime air as shown by temperature (*a*) and dew point (*b*) for Midway Island, Pacific Ocean; and equatorial air temperature (*c*) and dew point (*d*) over Albrook, Panama.

the low temperatures in the upper layers of polar continental and arctic air. If the circulation is such that the air remains over the northern ocean for two or three days, a state of equilibrium is approached. The air that reaches the Washington coast from the Pacific, or the west coast of the British Isles from the Atlantic (see Figure 8.5), is usually very moist and unstable in the lowest 2 km. At upper levels, it has a lapse rate close to the moist adiabatic, although much of the initial instability has disappeared. The result of these processes is polar maritime air.

When winter polar continental air is carried over open water such as the Sea of Japan or Lake Superior, modification takes place rapidly, but not for a long enough time to permit it to reach a stable temperature distribution. The modification does produce typical weather, that is, cumulus clouds and heavy snow showers on the lee shores, resulting in the snow belt along the Michigan shore of Lake Superior and elsewhere.

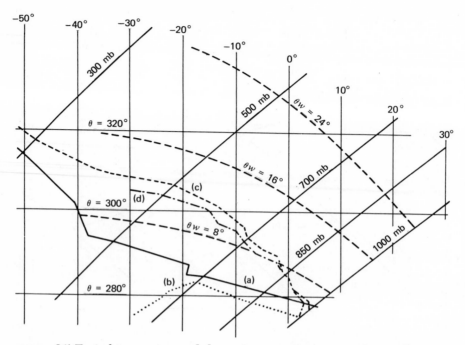

FIGURE 8.5 Typical temperature and dew point curves in winter with westerly winds carrying winter polar maritime air over Tatoosh, Washington (curves *a* and *b*) and Hannover, Germany (curves *c* and *d*).

In summer over the continents, evapotranspiration can be rapid from the water surfaces and the growing vegetation. These sources provide moisture that recondenses in clouds in the upper portions of some of the thermals which develop. Frequently, the vertical currents reach only low altitudes, and we get the familiar cumulus of fine weather. When, because of additional heating or because of advection of cold air aloft, the instability becomes great, air-mass thunderstorms develop. The evaporation from the underlying surface is a major reason for a difference between polar continental air in summer and that which develops over the tropical desert areas. Figure 8.6 gives examples of summer polar continental air.

8.5 Turbulence

The vertical eddies that form in the atmosphere mix the various layers and, with mixing, tend to produce a uniformity of the different conservative properties in the vertical. In a completely mixed layer, the mixing ratios of such properties as water, dust particles, and impurities are the same at all levels. This is not true of temperature, but is true of potential temperature with the result that the lapse rate becomes the adiabatic lapse rate.

Turbulence, therefore, is a process that tends to modify the characteristics of an air column. *Thermal turbulence* arises because of the instability of the air column and increases with increasing instability. *Mechanical turbulence* develops because of irregularities in the underlying surface or, in the free atmosphere, at a layer where there is a rapid change of wind with height. The rate of modification of an air column through turbulence depends on the intensity of the turbulence. Over either rough or heated ground, turbulent mixing proceeds rapidly, and the air column tends to develop a uniform moisture content with an adiabatic lapse rate. With a stable layer and light winds, the processes are slow.

When a layer of air is completely mixed, the top of the layer may be higher than the lifting condensation level of the surface air. At this

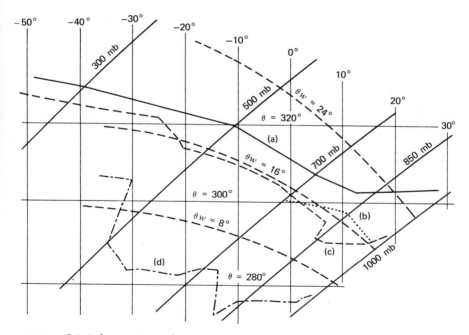

FIGURE 8.6 Polar continental air, temperature and dew point, in summer, typical afternoon conditions: Curves (*a*) and (*b*) for St. Cloud, Minnesota; and curves (*c*) and (*d*) for Fort Smith, Northwest Territories.

level a cloud forms, and above this level the temperature decreases at the moist adiabatic lapse rate. Some stratocumulus clouds over the ocean and a stratus cloud over open water in the polar regions form through turbulence under a stable layer aloft.

Two types of air mass develop rapidly through turbulent mixing. Polar continental air in winter rapidly changes to polar maritime air, and an air column that moves over a hot land area acquires uniform characteristics. In both cases, the underlying surface is warmer than the surface air. When the flow of air carries warm air over cooler land, turbulence is less intense and the modification through turbulence is slow, although not completely absent.

8.6 Air Masses, Weather, and Climate

Air masses in their source regions have typical characteristics that are reflected in the weather of these regions. There are variations of these characteristics through the year, modifying the weather. For any given locality within or outside a source region, the air column over the district has come along a path that the weather map can show. Along this path, the air column has been modified by wind shear, vertical motion, radiation, conduction, and evaporation. A consideration of the modifications permits one to estimate their effects on the air column and, therefore, on the resulting weather.

As an example, consider the flow of air from the south across the Gulf Coast. In midsummer, the land is hot and moist. The modification of the air above this land will increase the instability of the column to produce heavy cumulus clouds and showers. With the same flow but in the fall or early winter the effect of the underlying surface is to cool the surface layers of the air to develop an inversion. This inversion will restrict the upward movement of pollutants being released into the atmosphere. The area covered by the tropical air then will have fog, stratus cloud, and haze.

The weather maps provide many examples of air flows that can be examined to determine the modifying effects of the underlying surface on the air column and on the weather. The problems at the end of the chapter give some examples of these flows. It is impossible to consider all the possibilities, but certain ones are repeated frequently, helping to explain the basic weather or the climate of a district. The most common flows are associated with the general circulation, to be discussed in the next chapter. For this reason, a knowledge of this circulation is fundamental to an understanding of the climates of the world.

8.7 Air Mass Characteristics: Summary

TABLE 8.1 Air Mass Characteristics in Their Source Regions

Type of Air Mass	Source Regions	Winter or Summer	Temperature	Humidity	Stability
Arctic	Polar ocean	W	Very cold	High below inversion	Very stable
		S	Cold	High below inversion	Stable
Polar continental	Alaska, Northern Canada, Siberia	W	Very cold	High below inversion	Very stable
		S	Warm	Variable, frequently high	Conditional instability
Polar maritime	Oceans north of 45°	W	Mild	High, particularly near surface	Convectively unstable
		S	Cool	High	Slight conditional instability
Tropical continental	Sahara, Arabia	S	Hot	Low	Afternoon absolute instability
Tropical maritime	Central portion subtropical high		Warm	Dry above 1 km, moist below	Stable
Equatorial	Equatorial trough		Hot	Moist to high levels	Conditional instability

The characteristics of an air column are always being adjusted to the underlying surface. Over the source regions, these characteristics approach an equilibrium that may be considered to be typical of that air mass. Some of them have been discussed in the preceding sections. Table 8.1 summarizes the major characteristics.

Air columns, as they leave their source regions, become modified rapidly or slowly depending on the conditions. Also, during the intermediate seasons of spring and autumn the source regions of the polar air masses are changing. In considering air masses and air mass weather, we must recognize that the "typical" air mass is somewhat of an ideal, and we must be alert to the variations from the typical conditions that are listed in Table 8.1.

PROBLEMS AND EXERCISES

1. The wind aloft increases in speed with height but maintains the same direction. What is the effect on the stability of the air above a station? Explain.

2. Hudson Bay freezes over in late November and December. What effect will this have on the modification of polar continental air moving off its west coast? What will be the effect on the weather along its east coast?

3. During the growing season, hot dry air from the southwestern United States is carried across the Rockies to eastern Colorado and Kansas. How will it be modified? What will be the effect on the crops?

4. A summer southwest wind carries tropical maritime air northeastward parallel to the Atlantic coast of the United States and over the Labrador current. What will be the change in the air column and the resultant weather over the Grand Banks of Newfoundland?

5. Westerly winds in winter carry polar maritime air over the Washington coast, across the Rockies into Montana where it displaces polar continental air, and then eastward to North Dakota. What modifying effects will there be on the air mass, and in what manner will the typical air column change?

6. Winter polar maritime air is carried from the Atlantic across the Low Countries of Europe, then over northern Germany and Poland. How is the air modified in this path? Compare your answer with that for Problem 5.

7. Hot Sahara air is carried northeastward over the Mediterranean. What will be the characteristics of the air mass as it reaches the coast of Italy?

8. Explain, on the basis of Figure 3.11, why the eastern and western coasts of an ocean at latitude 25–40° will have different air-mass characteristics and, therefore, different weather. To what extent does this help to explain the difference in weather between northwestern Africa and the southeastern United States? Between California and the Carolinas?

9. Data from five radiosonde ascents are given in Problem 4, Chapter 6. What characteristics of these ascents give evidence of the environments in which they took place? How would you describe the air masses in which the ascents were taken?

CHAPTER 9

THE GENERAL CIRCULATION

9.1 *Thermally Driven Circulations*

In Chapter 1, the problem was raised concerning the manner by which heat is carried from the equator to the poles. It was suggested that the rotation of the earth prevented a direct flow from equator to pole and back. This subject was discussed at some length in Chapter 7, where it was shown that the air tends to move parallel to the isobars and not in the direction of the pressure gradient force.

Consider the movement of water in a large pan when heat is applied at the center. A flow establishes itself in which the water moves upward at the center, toward the sides at the top, downward near the sides, and back to the center of the dish along the bottom. This flow has many similarities to the sea breeze circulation, with pressure surfaces moving vertically (see Section 7.12). An example of a flow of this nature occurs over tropical islands where the solar radiation absorbed by the island corresponds to the heat source of the pan. The vertical currents are frequently topped by cumulus clouds, a familiar sight to sailors who thus are able to locate islands from afar. The horizontal convergence caused by the sea breezes around the Florida peninsula in summer accentuates the vertical currents over the land, a contributing factor to the prevalence of thunderstorms over the state.

If we could imagine that the earth was stationary and that the sun, our source of heat, revolved around the earth, the circulation within the atmosphere would not be greatly different from the circulation in the pan. Heated air near the equator would rise and, then, would move poleward because of a pressure gradient. Subsidence would take place in the polar regions, and the air would return toward the equator near the earth. This would be another example of a thermally driven current, in this instance extending over a wide area. But, as explained in Sections 7.4 and 7.5, the Coriolis force interferes with this flow.

9.2 *Dishpan Experiments*

During the past 20 years, laboratory experiments have helped toward an understanding of the flow of air over the earth. The pan of Section 9.1 has been rotated at various speeds to give rise to a force that corresponds to the Coriolis force. The heat is applied at the rim, and cooling at the center of the pan, to simulate a hot equator and a cold polar region.

When the pan is not rotated, the flow that becomes established is the simple thermally driven one. Even when the pan is rotated slowly this flow does not disappear, but the warm water at the top acquires a "westerly" component, and the water near the bottom drifts counter to the direction of rotation. In reality, it rotates in the same direction as the pan, but relative to the pan it flows backward. A "Coriolis force" has begun to act.

When the rate of rotation is very great, the flow breaks up into small, individual cells that come and go rapidly. With an intermediate rate of rotation, the flow takes on characteristics of the flow over the earth's surface. Anticyclonic flow appears near the bottom at about one third the distance from rim to center. This flow appears very stable. Nearer the center appear cyclonic whirls along a line of juxtaposition of two flows—one from the anticyclonic cells near the rim and one from the center of the pan. These cyclonic cells drift inward but are barred from the center by the outward-directed flow. At the top surface of the water, the flow is generally in the direction of rotation, but waves become established in this flow very similar to the waves found in the earth's troposphere above 5 km.

The total flow pattern provides a complex method of transferring warm water inward to the center and cold water outward to the rim of the pan. Some of the water transport is done within the eddies that are present near the bottom. The sinusoidal flow near the top of the water also provides considerable radial flow of water, both inward and outward. This total picture is closely related to the flow of the atmosphere over the earth's surface.

9.3 Major Features of the Earth's Flow

The features observed in the dishpan experiments have many similarities to the flow found on the surface of the earth. Reference should be made to Figures 3.10 and 3.11 that show the normal sea-level pressures for the months of January and July, respectively.

The sea-level trough of low pressure, produced by solar heating, is found near the equator on both maps. This trough has an annual movement northward and southward, following the sun. The thermally driven currents carry air upward in the vicinity of the surface trough, and poleward at high levels. These currents, coming under the influence of a stronger Coriolis force poleward of 15° lat, acquire westerly components. This results in convergence at high levels, subsidence, and belts of high pressure which ring the earth in the tropical regions. The return current at the surface at these latitudes forms the trade winds, veering somewhat toward the west because of the effect of the earth's rotation.

In the Southern Hemisphere, the Antarctic ice cap has a large net loss of heat, cooling the air over its surface. This air flows by gravity northward off the margin of the continent, being replaced by warmer air that moves in aloft. The outward current acquires heat from the unfrozen ocean and presently meets the warmer air that flows from the tropical high-pressure area. Along the line where cold and warm air meet, cyclonic circulations develop in which the warm air is lifted to be carried over the ice cap. Here, then, we have another thermally driven circulation. Notice in Figures 3.10 and 3.11 the persistent trough of low pressure at the edge of the ice cap, a trough that circles the polar continent. It is along this trough that the cold air from the ice cap and the warm air from the north meet.

The variation in the underlying surface around the North Pole makes the flow less regular than the one around the South Pole. There is in winter a large net loss of heat over Greenland, over the ice-covered Arctic Basin, and over the snow-covered areas of North America and Eurasia. Here are developed the polar continental and arctic air masses described in Section 8.3. Air masses moving outward from these areas meet the warmer flows from the south to produce the surface low-pressure centers and the jet streams of high levels. Vertical flows develop,

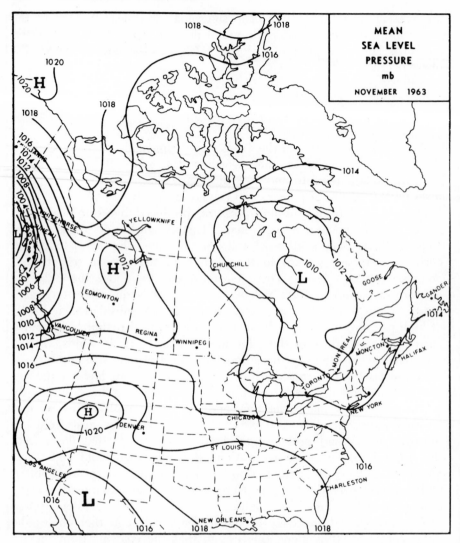

FIGURE 9.1 Mean sea-level pressure, November 1963. (This diagram and also Figures 9.2 to 9.6 are published, courtesy, Meteorological Service of Canada.)

and the circulation aloft carries warm air northward above the polar regions. The location of these thermally driven currents changes from day to day. The sources of cold air in summer are much smaller in area than in the winter season.

In both hemispheres, the cyclones that form at the boundaries between the cold air and the warm air reach their maximum intensity in the

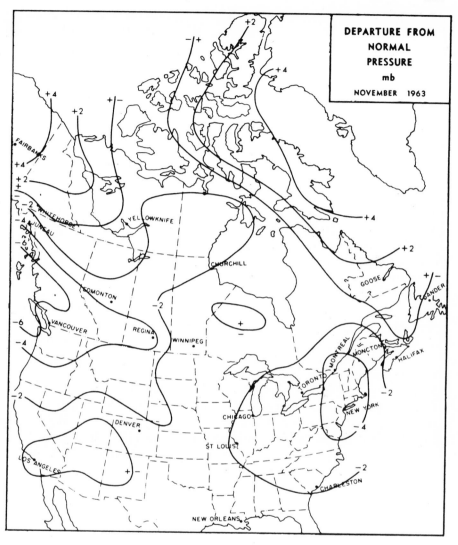

FIGURE 9.2 Departures from mean pressure, November 1963.

latitude band between 50° and 60°. Some, particularly in the Northern Hemisphere, move nearer the poles, but usually with decreasing intensity.

The flow of air between the subpolar lows at 60° and the tropical anticyclones at 30° is not well shown on the mean maps. This is the region of westerlies developed by the pressure gradient between belts of high and low pressure near the surface and accentuated aloft by the effect of the thermal winds (see Section 7.10). It is an area of traveling surface cyclones and anticyclones and of upper-level troughs and ridges in the wind stream that change their positions from day to day and from season to season.

In the averaging process to produce the mean-pressure charts for the Southern Hemisphere, these features disappear, and mean isobars tend to follow lines of latitude. In the Northern Hemisphere, the contrast between ocean and land interferes with the smooth west-to-east flow, so that mean maps can show some effects of these moving centers. These traveling surface centers produce a transport of warm air poleward and cold air equatorward. On the east side of every moving anticyclone, winds are blowing toward the equator, carrying cold air to regions where it can pick up warmth and moisture. This air returns poleward on the west side, modified and bearing energy both as sensible heat and latent heat of evaporation.

To the inhabitants of any one area, these flows produce an irregular cycle of cold and warmth. In times of active movement of the centers, a three-day cycle may persist for several weeks. Many weather proverbs are based on observations by laymen that, during the summer, storms return at approximately weekly intervals. But, at times, the pressure centers may stagnate and, for a period of two or three weeks, the weather may be abnormally cold, warm, wet, or dry.

Examples of the effect of stagnation are shown in the mean maps for November and December 1963. Figure 9.1 shows for North America the mean sea-level pressure for November, Figure 9.2 the departures from mean pressure, and Figure 9.3 the temperature departures from normal. Figures 9.4, 9.5, and 9.6, respectively, give the same maps for December 1963.

In November 1963, lows stagnated off the Alaskan panhandle, so that the mean pressure there was 6 mb below normal. This, combined with a high over Utah and Nevada, produced a strong westerly flow that brought air from the Pacific. across northwestern United States. This flow influenced most of the continental United States to give temperatures above normal except in California and Nevada. The maximum departure was 8°F in the vicinity of Lake Superior. To the northeast, a persistent low over western Quebec and higher pressures over Greenland and Baffin Bay kept cold air from sweeping over the eastern Canadian Archipelago, and here again the month was abnormally warm, 8°F above normal. In the northwest, persistent highs over Alaska and Yukon, producing polar continental air, resulted in temperatures as much as 14°F below normal.

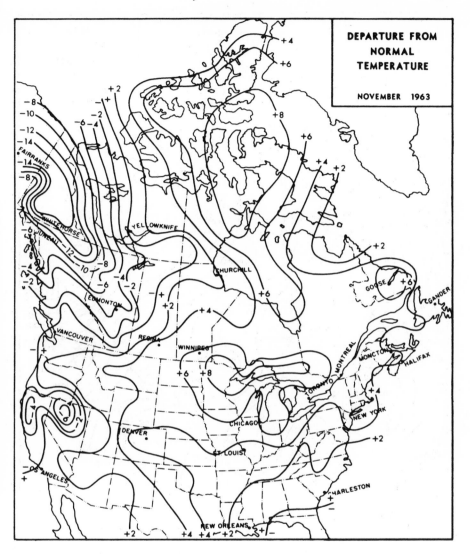

FIGURE 9.3 Departures from mean temperature (°F), November 1963.

December 1963 showed an almost complete reversal. A ridge along the 115°W meridian was much stronger and more persistent than normal. A strong flow to the west of it carried warm air into the Yukon and Alaska where the mean for the month was as much as 16°F above normal. To the east, the low off Labrador was 6 mb deeper than normal.

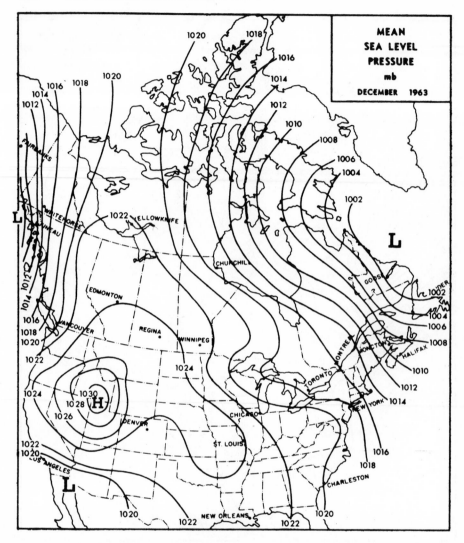

FIGURE 9.4 Mean sea-level pressure, December 1963.

The flow between the strong ridge at 115°W and the deep low off Labrador poured very cold air during the month over eastern Canada and the United States and even across the mountains to California. Two large areas, one centered in Illinois, the other along the lower St. Lawrence, had mean temperatures 10 deg F below normal.

In neither month was the wind flow constant. The pressure centers

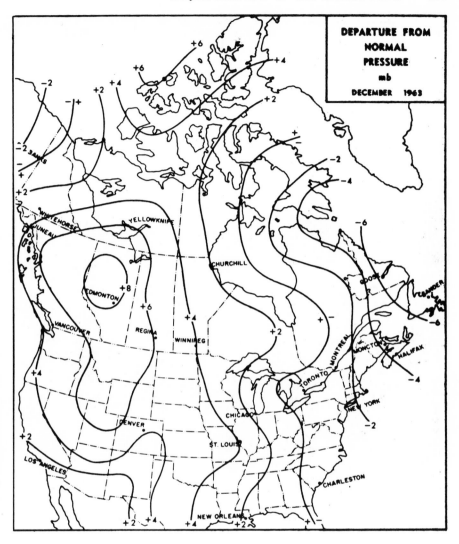

FIGURE 9.5 Departures from mean pressure, December 1963.

moved, but there was a tendency for them to reestablish themselves in certain areas—a tendency that resulted in extreme departures from normal. Meteorologists are beginning to understand some of the causes of the general circulation, but they have a long way to go to understand the reasons why two months can differ so widely as November and December, 1963.

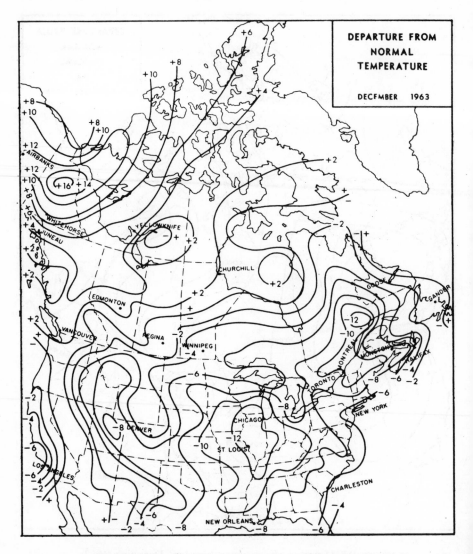

FIGURE 9.6 Departures from mean temperature (°F), December 1963.

9.4 *Land and Ocean*

Land and sea breezes (discussed in Section 7.12) are thermally driven circulations caused by the temperature differences between coastal areas and their adjacent waters, circulations with a 24-hour rhythm. Section 9.3 discussed the thermally driven circulation of the atmosphere caused by the heating of equatorial regions in contrast to the polar areas. This circulation persists, although with some change in strength, throughout the year. A thermally driven circulation intermediate in size between these two is the circulation caused by the differential heating between continents and oceans. The phenomenon shows an annual variation related to the solar heating.

In all three circulations, the driving forces are pressure gradient forces. With land and sea breezes, the pressure differences are small and difficult to detect. Considering the annual cycle between land and ocean, the effects of the temperature difference on the pressure gradients can be observed if averages for sufficiently long periods are taken. Notice the difference in the mean pressure patterns over the Great Lakes for November and for August that are shown, respectively, in Figures 9.7 and 9.8. In November, these lakes are unfrozen and heat is supplied to the air that passes over them. This tends to make the pressures relatively low. In contrast, in summer the lakes remain comparatively cool, contributing to a high pressure over the water areas. These developments are the cause of the wind reversal (see Section 7.12) of offshore winds in winter and onshore winds in summer.

The effects of the lakes on the pressure distribution do not show up clearly on individual maps. As in other areas of the temperate zone, the lakes have their succession of high-pressure areas and low-pressure areas. In the succession, high-pressure centers become slightly more intense when the lakes are relatively cool, and low-pressure centers deeper when the lakes are warm. The rapid development and intensity of November storms on the three largest lakes are well known to people who travel over these bodies of water.

The effect of the annual temperature cycle on the pressure cycle becomes greater as we consider larger bodies of land or water. Australia provides a good example of the effects of the rhythm. Latitudinally, the continent is in the zone of the subtropical high-pressure areas, but in summer (Figure 9.9), with an overhead sun, the heat is excessive. The resulting low pressure attracts unto itself the equatorial trough which now lies over the northern half of the continent. In winter (Figure 9.10), the relatively cool land augments the normal high pressure of the subtropical belt, so that pressure is higher than it is over the adjacent ocean.

Another example is found in the Mediterranean area, an area of relative warmth during the winter season, and so a region of development and movement of many storms. The cool surface in summer produces a high-pressure area, augmenting the normal subtropical high-pressure belt. This results in light winds and clear skies.

The extreme example of the effect of land and ocean on the pressure distribution is found over Eurasia, the largest land mass on the earth.

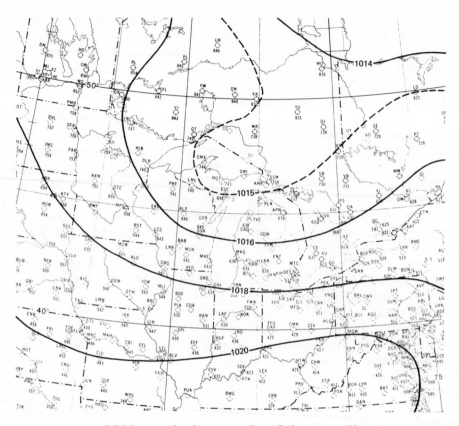

FIGURE 9.7 Mean sea-level pressure, Great Lakes region, November.

In winter, the cooling by radiation produces very low temperatures and correspondingly high pressures. Almost all weather maps of the winter seasons show a high-pressure center over Siberia. This in turn produces a clockwise movement of air, carrying cold air off the continent. This is particularly true on the east coast where the northwest monsoon brings cold air to low latitudes. South Korea at the latitude of northern California has a January mean temperature 25 deg F colder as a result of these winds. Over southeast Asia this wind has become the northeast monsoon. This is the dry monsoon of the Indian peninsula except where it approaches an eastward-facing shore. On these coasts, the winds carry moist air off the tropical water, often causing orographic rain.

Summer in Asia brings a reversal of the monsoon circulation (see Figure 3.11). Now the heating is excessive, particularly in the tropics. A low-pressure center is formed in the northwestern corner of the Indian peninsula to dominate the circulation. The equatorial trough moves northward to pass through this center, and about it the winds tend to have a counterclockwise motion. The southeast trades of the Southern Hemisphere cross the equator and, deflected by the influence of the Coriolis force of the Northern Hemisphere, become the southwest trades of the Indian Ocean and south Asia. Counterclockwise winds along the east coast carry moist tropical air over the coast of China and the off-shore islands.

The day-to-day weather maps do not, of course, always show the configuration of the mean maps of Figures 3.10 and 3.11. For Asia the deviations from the mean are less than for other areas in the region of westerlies, yet even here differences occur. The winter high sometimes is as far west as European Russia, and sometimes is split into two parts. The summer monsoons at times increase in strength, at times weaken.

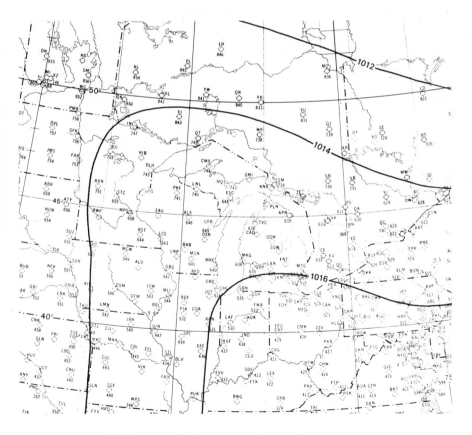

FIGURE 9.8 Mean sea-level pressure, Great Lakes region, August.

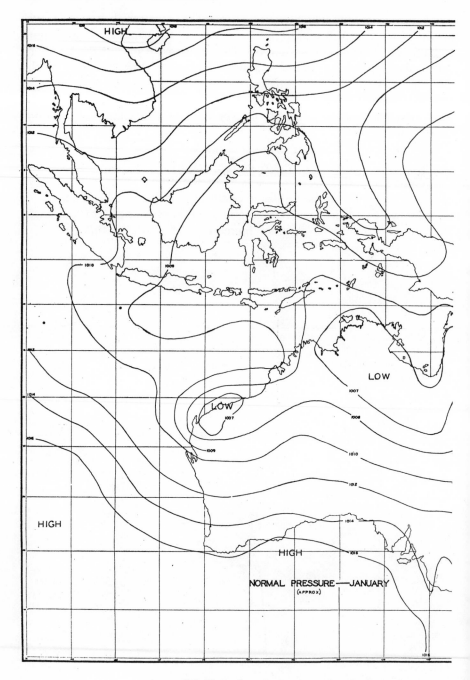

FIGURE 9.9 Normal pressure, Australia, January. (Courtesy,

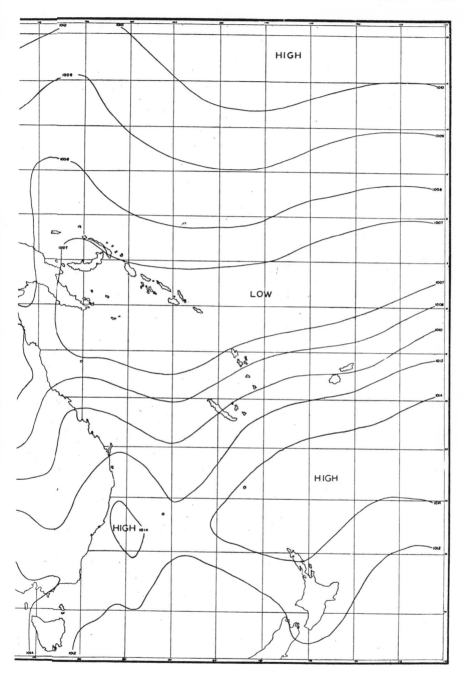

Bureau of Meteorology, Commonwealth of Australia.)

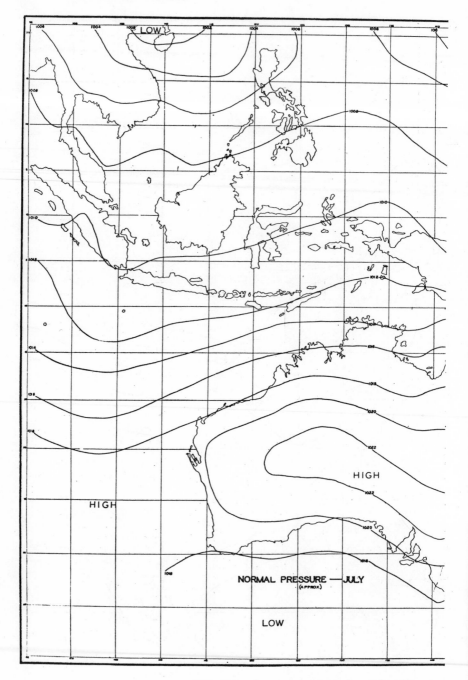

FIGURE 9.10 Normal pressure, Australia, July. (Courtesy,

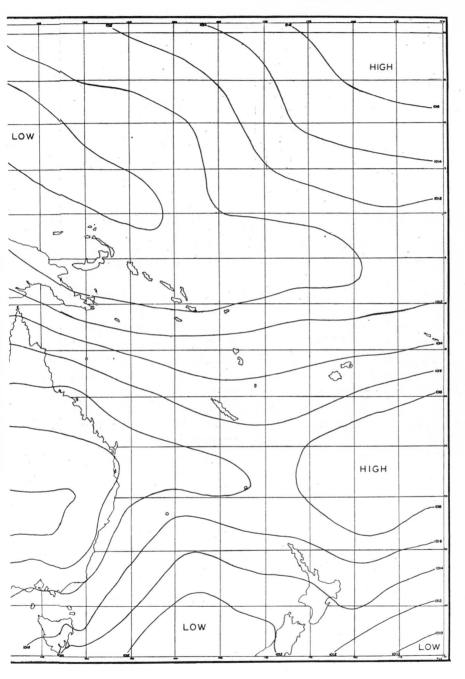

Bureau of Meteorology, Commonwealth of Australia.)

But the normal pattern of flow persists, and as such provides meridional flows that carry cold air southward and warm air northward.

In North America the effects of the seasons on the pressure pattern are significant, although not so pronounced as in Eurasia. In general, the pressure over the continent is high in winter, with a ridge extending northward along the Mackenzie River to the Arctic. This causes offshore winds along the east coast of the continent. In summer the low-pressure area causes a monsoonal circulation, although not so strong as the Indian monsoon, from the Gulf of Mexico and the tropical Atlantic over the southeastern sections of the continent. From this flow comes much of the moisture that falls on the United States during the summer months.

9.5 *The General Circulation and Poleward Transfer of Heat*

Experiments such as the dishpan experiment and computations using electronic computers are helping to solve the problems of the general circulation. Much remains to be done, but with the help of the discussion of the preceding pages, we can see the general outline of the pattern by which equatorial heat passes toward the poles.

Figure 9.11 shows the average cross section from equator to pole. It indicates the equatorial trough, the trade-wind belt, the tropical anticyclone, the region of westerlies, the subpolar low, and the polar high. These bands move north and south, following the sun. Superimposed on them are the land and sea effects, so that highs become more intense over land areas in winter, and over ocean areas in summer. The traveling cyclones and anticyclones of the region of westerlies form a second modification to the general weather picture.

The transfer of heat in the thermally driven circulation between the

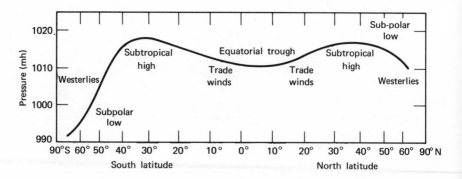

FIGURE 9.11 Mean pressure profile with latitude for the year.

equator and tropics presents no problem. The flow across the belt of high pressure at 25–40° latitude would be difficult to understand if the belt were continuous, even though some of the upper-level air, with its high potential temperatures derived in the equatorial zone, does move across the band into the region of the westerlies. However, the subtropical high tends to break itself up into cells. These cells are easily noted on weather maps for the ocean areas in summer but, even at other times of the year, the belts are not continuous. The large horizontal eddies formed by these cells transfer warm moist air poleward on their western edges and cool drier air equatorward along the eastern boundaries. For example, the summer monsoons of southern Asia and southeast United States carry much sensible and latent heat energy into the temperate zone. Meanwhile, cool air moves southward along the Pacific coast from Alaska to Oregon and California.

The daily weather maps provide the clue to the movement of heat energy across the zone of westerlies. Migrating cyclones and anticyclones form small or large eddies to provide meridional flow in these latitudes. The sinusoidal flow of the principal jet stream in the upper troposphere is also part of the overall flow. The poleward-moving branches of the flow are warmer and more moist than the equatorward-moving branches, so that these provide a significant heat transport.

Although surface eddies migrate, they tend to stagnate in certain locations. The lows of the North Pacific are often found near the Aleutians, those of the North Atlantic near Iceland, particularly in winter. They tend to disrupt the west-to-east mean isotherms. Southerly winds with their warm air tend to be located along the British Columbia and Alaskan coast and, in the Atlantic, along the west coast of Europe. Thus these areas are warm for their latitudes. Meanwhile, northerly winds along the east coasts of Asia and North America cause these areas to be relatively cool.

The snow-covered polar areas are the source regions of polar continental and arctic air. These air masses finally reach the stage where they break away from their source to be caught in the flow of the zone of westerlies. In this movement the air picks up heat, slowly or rapidly as the case may be, to replace that which had been lost. Meanwhile, the people in the path shiver in the cold arctic air. For reasons not yet fully understood, these flows tend to follow one another along the same path for a period of a week or longer. This location is governed somewhat by the positions of the troughs and ridges of the waves in the upper atmosphere. The area where the breakouts occur will have extremely cold weather, while other areas at the same latitude but in the path of the returning flow will experience a mild spell.

CHAPTER 10

WEATHER MAPS AND EXTRATROPICAL CYCLONES

10.1 *Construction of a Weather Map*

Basic to the study of weather is the weather map. References have been made to this tool of the meteorologists in the earlier chapters of this book. The production of a weather map gives an excellent example of international cooperation and of technical skills of communication. These are necessary, first, because weather processes do not recognize international boundaries and, second, because the value of a weather map for forecast purposes decreases rapidly with time. A 24-hour-old map has little value for forecast purposes except as a background to the current weather picture.

The weather map gives an instantaneous snapshot of a moving picture, a record of the weather at a specific time. By agreement among the nations of the world, weather observations are made regularly at 0000, 0600, 1200, and 1800 GMT, the *synoptic hours*—observations which give a complete picture of the current weather. From the recorded data, significant information is deduced or calculated, put into coded form,

204

and distributed to all parts of the world. Weather reports include information such as cloud cover, wind direction and speed, sea-level pressure and its change during the past three hours, temperature, dew point, current and past weather, visibility, cloud types and amounts of each type, and the amount of precipitation during the past six hours.

The weather reports from the synoptic stations are collected by radio, telegraph, and teletype, and then interchanged by a vast communications network to forecast offices over the hemisphere. Here the data are plotted on maps, again in special codes. Figure 10.1 gives a sample of the plotting for a particular station. Temperature is plotted at the upper left and dew point at the lower left, both in degrees Fahrenheit; at the upper right is the sea-level pressure in millibars and tenths with the hundreds figure omitted, and below it the pressure change during the past three hours, in millibars and tenths. The code for the present weather, plotted between temperature and dew point, is composed of some basic symbols that are, on the whole, fairly descriptive of the actual weather phenomena. Some of them are given in Table 10.1. The visibility in miles is given to the left of the station circle. With this background, it is possible for a layman to learn much of the significant information, although he will not be able to interpret all the detail.

A number of stations also observe and record the winds aloft by pilot balloons. A small group of stations make observations of temperature, humidity, and winds aloft by radiosonde balloons. The assistants also plot the upper-air data on emagrams or tephigrams, and significant data on the series of upper-air charts: 850, 700, 500 mb, etc.

The maps and charts are then passed to the forecasters who analyze them in various ways. On the upper-air charts are drawn the constant-height lines, which should parallel the winds at that level, and also the isotherms. The plotted ascents are used to assist in determining stability and fronts. These fronts are drawn on the weather charts, and

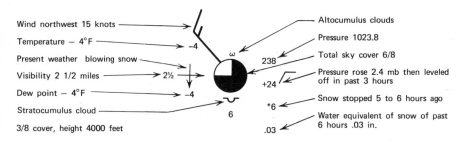

Wind northwest 15 knots

Temperature — 4°F

Present weather   blowing snow

Visibility 2 1/2 miles

Dew point — 4°F

Stratocumulus cloud

3/8 cover, height 4000 feet

Altocumulus clouds

Pressure 1023.8

Total sky cover 6/8

Pressure rose 2.4 mb then leveled off in past 3 hours

Snow stopped 5 to 6 hours ago

Water equivalent of snow of past 6 hours .03 in.

FIGURE 10.1 Plotting model for weather observations.

TABLE 10.1 Frequently Used Symbols to Indicate Weather

Rain	·	Fog	≡	Haze	∞	Blowing snow	
Snow	*	Shower	▽	Thunderstorm	$\overline{<}$	Weather in sight	( )
Drizzle	'	Hail	△	Dust	S	Weather ending in past hour	( )

then isobars are drawn. In all this study, the preceding maps and charts are used in order to recognize and to understand the changes that are occurring.

The observation and collection of data have one major purpose—the forecasting of the weather as far into the future as possible. Before the advent of the computer, the analysis was done by the meteorologist in the manner outlined in the preceding paragraph. The complexity of the problem of forecasting and the pressure to release the forecast as soon as possible prevented him from assimilating completely the available data. The forecast map was prepared with considerable help from his experience with previous maps and with little use of the mathematics and physics on which his training had been based.

The computer has brought about a revolution. With the collected data, the computer solves numerically a number of differential equations to compute Δp, the change in pressure that will occur during a given time interval—for example, one hour. This is done for a large number of grid points. The result should give the pressure pattern for the end of the time interval. Now the computer takes these new data and again computes new values of Δp. This process, of course, can continue indefinitely.

The original work with computers was based on semihorizontal flow over uniform terrain. The results from the computer were found to agree reasonably well with the developments on the weather map over certain areas of the earth where the initial conditions were satisfied. In other regions significant departures from the weather map occurred. It has been found necessary to adjust the mathematical equations used by the computer to take into account the transfer of heat from or to the underlying surface, and also the effect of mountain ridges on the flow over them. With these and other minor changes in the basic conditions, the computer is now able to produce forecast maps for periods up to 48 hours which are better than the ones made by even an experienced forecaster. The computer can also project, with reasonable success, the maps up to five days, and forecasters are working toward longer periods.

Two deficiencies limit the efforts of the computer to go on indefinitely. First, the complexities of the flow of air over the surface of the earth cannot be inserted into any set of mathematical equations. The effects of approximations in the basic equations and of errors in the data increase with time to cause the forecast to differ from the actual development. The second deficiency is in the lack of basic data. In general, over the land areas of the earth, the density of radiosonde stations provides sufficient data, although many areas have too few stations. The oceans cover 75 per cent of the globe, and over them the density of stations is small. This is particularly true of the Southern Hemisphere. Islands help fill these blank spots and, in the north Atlantic and north Pacific, weather ships are stationed at specific locations to take observations. In spite of these efforts, the data are too scattered to be of maximum use to the computer.

Satellite data are being examined and methods devised to use them in the computer program. Already these data are being used to adjust the flow pattern at 500 mb and above, and further use will be made as experience with them grows. The World Weather Watch 1968 to 1971 of the World Meteorological Organization may be able to provide some missing but significant information and methods by which the data can be used more effectively. Balloons which rise to a specified level and there float with the winds are tracked by radar from stations around the globe to provide the winds at that level; temperatures also are available. Already, some balloons have circled the earth for a year at altitudes above 10 km in the region of westerlies in the Southern Hemisphere. The failure of others has revealed problems that must be overcome.

The two deficiencies—the lack of data and approximations in the basic mathematical equations—are being studied, and adjustments will permit improvements to be made in the forecast maps. Even knowing the changes that will occur, the meteorologists who are working on the problem recognize that errors cannot be eliminated. They foresee that the maximum time for which this method can be used is ten days or two weeks.

The use of electronic computers has not eliminated the practising forecaster. He is relieved of some tasks, but he must still examine the output from the computer to estimate the maximum temperature or the height of the cloud base or the intensity of fog. The computer is able to assist in some of these estimations because it is able to prepare maps that were impossible for the forecaster to do in the short time available. For instance, the computer is able to prepare a map showing the speed of vertical motion over an area, a quantity that, of course, is related to the extent and depth of cloud and the amount of precipitation to be expected. At present and for some time to come, the most accurate predictions are made by man and computer, that is, computer solutions modified by man on the basis of his experience and his knowledge of the approximations used in the mathematical models.

10.2 *Interpreting the Weather Map*

Fundamentally, the climate at any place is determined by the sun, but the day-to-day changes in weather are caused by the changing winds. Where the winds are steady, as in the area of the trade winds, day-to-day changes are small. Where changes of wind occur, the weather will also change.

The weather map is useful since it gives a clue to the winds and their variation. This is possible since the winds above the frictional level are governed by the pressure distribution. The spacing of isobars on a weather map is determined by the scale of the map and the horizontal pressure gradient. The 4-mb interval used in this text is the one most frequently used in the United States and Canada.

An example of a weather map, with samples of some typical patterns, is given in Figure 10.2, the weather map for eastern United States for 1200 h GMT 8 January 1969. The temperature and wind data for this map are given in Figure 2.9, and the pressure data in Figure 3.8.

At the time of the map, snow was reported in Quebec and New England, associated with a cyclone in New Brunswick. In New York State there were some snow showers from air that had been modified by Lake Ontario. A center of high pressure lay along the Georgia-South Carolina coast. The low temperatures and dew points provide evidence that this air was polar continental air which had recently moved southward from Canada.

The air over Texas was mild, in the fifties, and in eastern Texas and Louisiana the air, having come over the Gulf of Mexico, was moist. Fog was reported at Houston, Texas in this moist air. The northern extent of this mild air is marked by a heavy line with semicircles and points—a front, going from Mississippi westward and northward to Denver, Colorado. Another front from Iowa through South Dakota to Montana marks the southern edge of a fresh outbreak of polar continental air. Snow was falling in this fresh outbreak and eastward to Lake Superior.

10.3 *Changes in the Weather Map*

The weather is continually changing, and these changes are reflected in the changing pattern on the weather map. Some patterns of change are repeated with minor variations. For example, lows frequently move into the Gulf of Alaska and stagnate in that area. Other developments occur only rarely. The 1938 hurricane that hit the Connecticut coast was the first hurricane to do so in 50 years, although many had passed to the southeast of the area.

Some appreciation of the changes in a weather map can be obtained by comparing Figures 10.2, 10.3, and 10.4. These latter two diagrams give the weather maps for 0000 h GMT, and 1200 h GMT, 9 January 1969, respectively; that is, 12 and 24 hours later than the map shown in Figure 10.2.

During the 24 hours, the low near Denver filled so that by 0000 h GMT 9 January, an elongated low lay east of the Mississippi between Memphis, Tennessee and central Wisconsin. At 1200 h GMT, the low was located in the lower Great Lakes area. With this movement, the area of snow moved into the Great Lakes area and south to cover Pennsylvania. During the 24 hours some freezing rain fell in Illinois, Indiana, and Ohio.

West of the low, a ridge of high pressure built southward from the Canadian prairies as north winds brought very cold air into the area west of the Mississippi River. The warm moist air over eastern Texas was replaced by the colder air from the north, but strengthening south winds carried warm moist air off the Gulf and over most of Mississippi, Alabama, and Georgia. The high in the southeastern United States moved into the Atlantic, and the New Brunswick low moved eastward toward Newfoundland.

The movement of pressure centers from west to east shown in Figures 10.2, 10.3, and 10.4 is common. The rates of motion at this time were much higher than usual, but there is no "normal" speed. Pressure centers may move at speeds up to 2000 km (1200 mi) in a day. Others stagnate and remain almost stationary, or may at times move slowly westward. The winds above 3 km give some guidance to the direction and speed of motion, but there is so simple rule.

Further details of changes in the weather map are illustrated in Figures 10.5 and 10.6. These show, respectively, the locations of centers of anticyclones and cyclones that appeared during the month of December 1957 north of 20°N, as shown on the daily maps for 1200 h GMT published by the United States Weather Bureau. According to Figure 10.5, centers of high pressure were frequently present in the Atlantic and Pacific between 30° and 45°N. They were cells of the subtropical high-pressure belt. Figure 10.5 also shows that centers of high pressure developed over the land areas of China and eastern Siberia, Greenland, and the area east of the Rockies. A number of small highs were centered

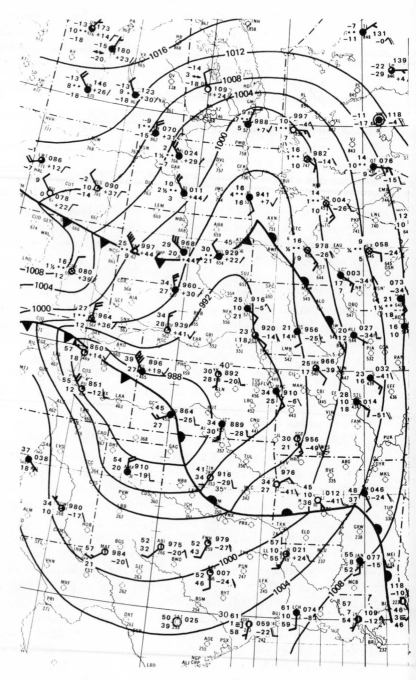

FIGURE 10.2 Weather map for eastern North

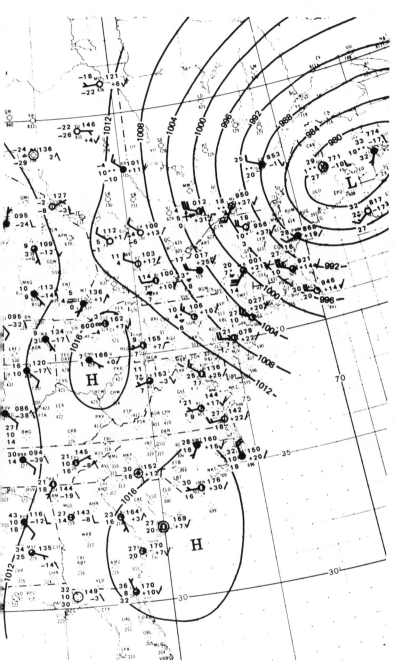

America, 1200 h GMT 8 January 1969.

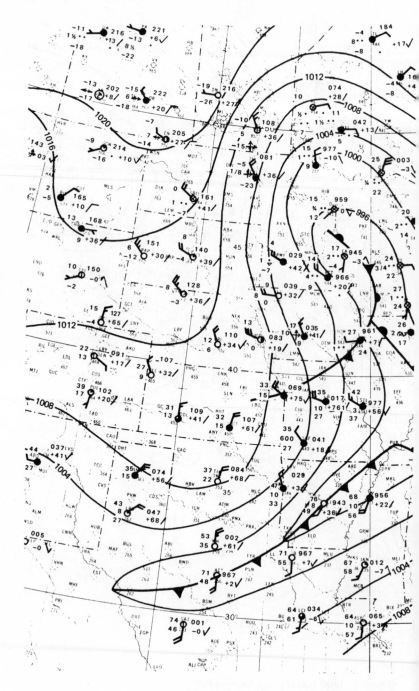

FIGURE 10.3 Weather map for eastern North

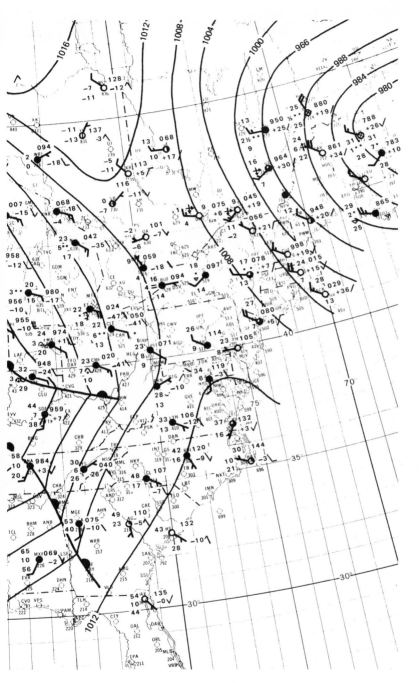

America, 0000 h GMT 9 January 1969.

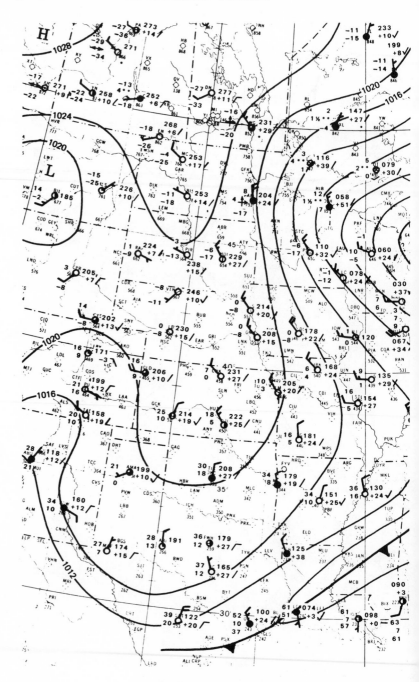

FIGURE 10.4 Weather map for eastern North

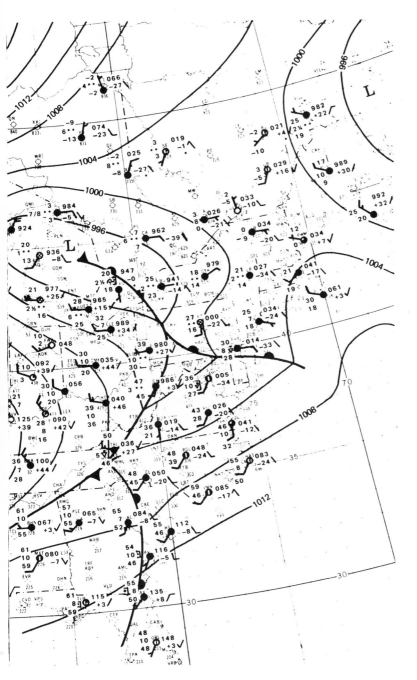

America, 1200 h GMT 9 January 1969.

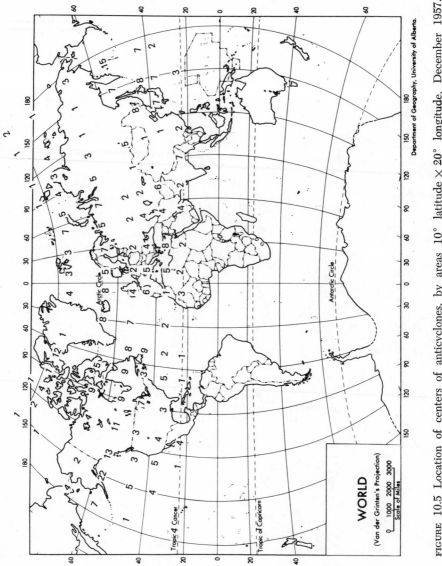

FIGURE 10.5 Location of centers of anticyclones, by areas 10° latitude × 20° longitude, December 1957.

Department of Geography, University of Alberta.

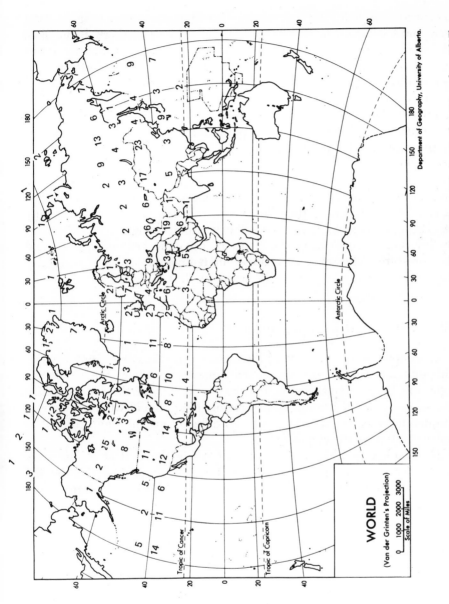

FIGURE 10.6 Location of centers of cyclones by areas 10° latitude × 20° longitude, December 1957.

in central Europe during that month. Low-pressure centers (Figure 10.6) were most common in the Gulf of Alaska and in northern Baffin Bay. The centers associated with the Icelandic low were not concentrated but were distributed from Newfoundland to the Arctic Ocean north of eastern Europe. The waters off the east coast of Asia were often visited by storms that were moving toward the Aleutian low. The numerous storms of the Canadian prairies were offshoots of this same area of low pressure. The most common place for storms south of 40°N was in the eastern Mediterranean, a result of the relative warmth of the underlying water.

In summer, the semipermanent anticyclones over the oceans are more prominent than in winter. Centers of low pressure are not so well marked as in winter, and usually their life cycles are much shorter. In the vicinity of a semipermanent high or low, the weather changes little from day to day. Other districts, in the paths of traveling centers of high and low pressure, have marked changes in temperature from day to day with periods of rainy weather alternating with periods of clear skies.

10.4 *Fronts and Frontal Surfaces*

In Chapter 8, the concept of air masses was discussed. They usually, although not always, develop in an anticyclone where the winds are light and the air remains over the district for two or three days. When the air moves away from its source region, the different wind circulations frequently bring together air masses of different origin. Figure 10.2 illustrates this type of development.

The south winds blowing across the coast of the Gulf of Mexico west of the Mississippi were bringing air that had been modified by the Gulf, and farther west the air had become warmed over the southwestern United States. Generally, this air had temperatures in the 50's. North of this air, over Nebraska and neighboring regions, the slightly modified polar continental air had temperatures in the 20's and 30's. The boundary line between the two air masses extended from the Gulf coast to Denver, Colorado. At the time of the map, there was little significant change across the front except for temperature and a wind shift. From western Kansas eastward, the flow was such as to cause warm air to replace cold, a major criterion for a warm front. This portion is marked with semicircles. West of the center of low, cold air was replacing warm, and we have a cold front, marked by points. In the region of the cold front, pressures were rising rapidly north of the front and falling south of the front.

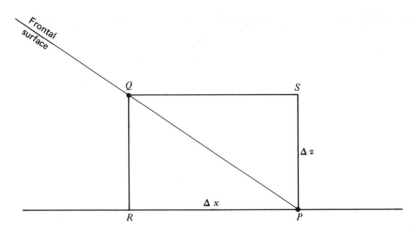

FIGURE 10.7 Slope of a frontal surface.

Another front lay north of the first front, in this case lying between modified polar continental air and a fresh outbreak of arctic air. Some snow was falling north of this front, partly because the warmer air was being forced over the colder air and partly through instability in the cold air as it moved over warmer ground. The warm front between these two air masses extended from western Minnesota to southeast Iowa, and the cold front west through South Dakota and Montana.

When cold and warm water lie side by side, there may be some mixing, but the warm water will tend to move above the cold. The same tendency exists in the atmosphere, but the boundary surface between the masses of warm and cold air is not level. Consider Figure 10.7 where P is the surface front with cold air to the left and warm air to the right. Assume that the bounding surface between the two masses is along PQ. Let $PRQS$ be a rectangle with PQ the diagonal and PR along the earth's surface. Let $PR = \Delta x$ and $PS = \Delta z$; also let p_P, p_Q, p_R and p_S be, respectively, the pressures at the four corners of the rectangle. Now

$$p_Q - p_P = p_Q - p_R + p_R - p_P = \left(\frac{\Delta p}{\Delta z}\right)_c \Delta z + \left(\frac{\Delta p}{\Delta x}\right)_c \Delta x$$

where the subscript c indicates that we are measuring pressure gradients in the cold air. Similarly

$$p_Q - p_P = \left(\frac{\Delta p}{\Delta z}\right)_w \Delta z + \left(\frac{\Delta p}{\Delta x}\right)_w \Delta x$$

where gradients are measured in the warm air as indicated by the subscripts. Since p_P and p_Q do not change as we cross the frontal surface,

$$\left(\frac{\Delta p}{\Delta z}\right)_w \Delta z + \left(\frac{\Delta p}{\Delta x}\right)_w \Delta x = \left(\frac{\Delta p}{\Delta z}\right)_c \Delta z + \left(\frac{\Delta p}{\Delta x}\right)_c \Delta x$$

Solving,

$$\frac{\Delta z}{\Delta x} = \frac{(\Delta p/\Delta x)_c - (\Delta p/\Delta x)_w}{(\Delta p/\Delta z)_w - (\Delta p/\Delta z)_c}$$

We may substitute in the denominator from the hydrostatic equation, and in the numerator from the geostrophic wind equation. When, as in this instance, the change in pressure is measured along a specific line, the geostrophic wind derived from the equation is that component normal to the line. In this instance, the wind will be the component parallel to the front. Therefore

$$\frac{\Delta z}{\Delta x} = \frac{\rho_c 2\Omega \sin \phi v_c - \rho_w 2\Omega \sin \phi v_w}{-\rho_w g + \rho_c g}$$

$$= \frac{2\Omega \sin \phi}{g} \cdot \frac{\rho_c v_c - \rho_w v_w}{\rho_c - \rho_w}$$

Once again v and ρ are, respectively, the symbols for velocity and density, g represents the acceleration of gravity, and $2\Omega \sin \phi$ is the Coriolis parameter. Substitution for density from the gas law, Equation 4, Chapter 3, gives

$$\frac{\Delta z}{\Delta x} = \frac{2\Omega \sin \phi}{g} \cdot \frac{T_w v_c - T_c v_w}{T_w - T_c}$$

$$= \frac{2\Omega \sin \phi}{g} \frac{\overline{T}(v_c - v_w)}{\Delta T} \tag{1}$$

where $\overline{T}$ is a mean temperature. This equation gives the slope of the frontal surface in terms of the temperatures in the two air masses and the difference between the geostrophic wind components parallel to the front. This latter quantity is obtained by determining $\Delta p/\Delta x$ along a line perpendicular to the front.

An approximate value to the slope of a front may be taken by inserting values into Equation 1. If we take $\phi = 35°$, $\overline{T} = 280°K$, $\Delta T = 6°$, and $v_c - v_w = 10$ m sec^{-1}, $\Delta z/\Delta x = \frac{1}{250}$. Figures 10.8 and 10.9 show cross sections of fronts as determined from temperature data. They give slopes ranging from $\frac{1}{100}$ to $\frac{1}{250}$.

The slope of a front may appear unimportant. Along a road, slopes of these magnitudes would not be noticed. But in the atmosphere these slopes become significant. As shown in Figure 10.9, the warm air that

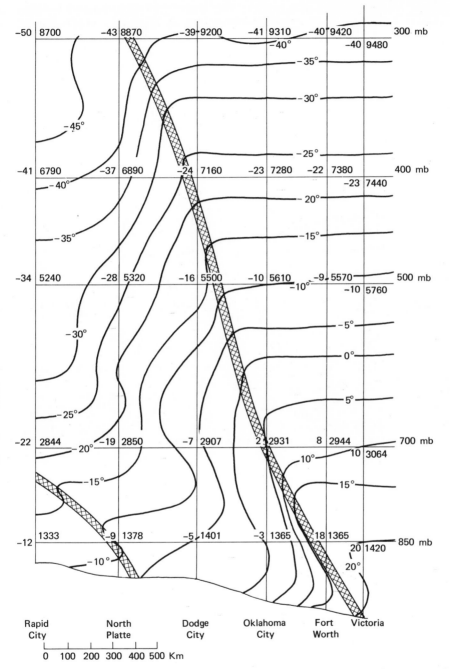

FIGURE 10.8 Cross section, Victoria, Texas to Rapid City, South Dakota, 0000 h GMT 9 January 1969.

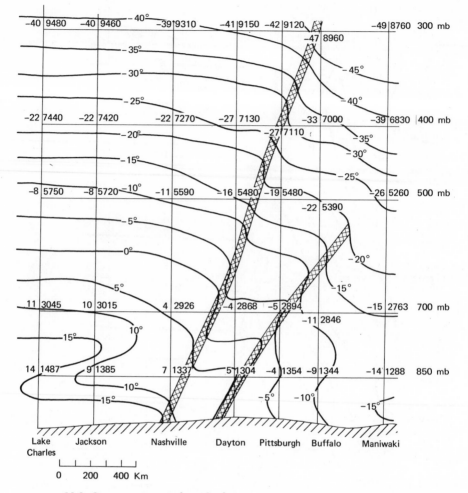

FIGURE 10.9 Cross section, Lake Charles, Louisiana to Maniwaki, Quebec, 0000 h GMT 9 January 1969.

was leaving the ground at Nashville, Tennessee on 9 January, 0000 h GMT, reached Pittsburgh 7 km above the ground. By this time, it had long passed its lifting condensation level, and some of the water had turned to ice crystals and fallen to the ground as snow. Flows above frontal surfaces do much to determine the pattern of precipitation on the ground.

In general, a *frontal surface* is the bounding surface between two air masses with different source regions and different characteristics.

The *front* is the intersection of such a frontal surface with the ground. On the scale of a weather map, the boundary is marked by a line, but in actual weather situations the temperature change from warm air to cold is gradual. On 8 January 1969 in the Oklahoma region, it changed about 10 deg C in 300 km. In some weather situations, the mixing of air near the earth makes it difficult to locate the boundary precisely, and it becomes necessary to examine the upper-air temperatures to locate the front. As a radiosonde balloon passes from the cold air through the surface into the warm air, the potential temperature and wet-bulb potential temperature rise rapidly. These conservative properties are more useful than temperature in locating the lower boundary of the warm air.

Help in locating and in examining the characteristics of a front is obtained by drawing a cross section through the front. Figure 10.8 shows the cross section from Victoria, Texas north to Rapid City, South Dakota. At this time, the major front (a cold front) was near Victoria, Texas at the surface, and above Dodge City lay just below the 400-mb surface, at approximately 7 km. Notice the uniform temperature in the horizontal in the warm air. In the cold air, modification by moving over warmer ground and also by heating from the warm air aloft caused it to become warmer near the front, but the temperature dropped rapidly in a horizontal line away from the frontal surface.

Figure 10.9 shows the cross section from Lake Charles, Louisiana to Maniwaki, Quebec, a line that intersected both fronts. As with the cold front, the temperature dropped rapidly along a horizontal line through the frontal surface. Of particular interest is the 0°C isotherm. In the warm air, this lay at 4.5 km, and in the modified polar continental air at 2 km. Surface temperatures were below freezing in the arctic air. At Dayton, Ohio, the surface, in arctic air, was below freezing, but at 850 mb, 1.3 km, in polar continental air, the temperature was 5°C. This temperature distribution resulted in the freezing rain that fell on Ohio at that time.

Because of the nature of a front, the isotherms for any level are closely packed and parallel to the boundary surface. As described in Section 7.10, this gives rise to a thermal wind—a wind shift with height that tends to make the winds parallel to the isotherms and therefore the front. Figure 10.8 helps to illustrate this effect. According to Figure 10.3, at 0000 h GMT 9 January the pressure at Rapid City was 1015 mb, and there was a gradual drop to 996 mb in Texas. With such a pressure gradient, the geostrophic wind was easterly, with a mean value of 11 m sec^{-1}. Aloft, the temperature contrast changed the situation. The slope of the 850-mb surface, although slight, was toward the north, and the height at Rapid City was 87 m below that at Victoria. The height of the 300-mb surface was 9480 m at Victoria, and 8700 m at Rapid City, or a slope of 780 m in 1600 km. The mean geostrophic wind was west and, based on a latitude of 38°, 53 m sec⁻¹. This was in the region of the jet stream. A careful analysis of the 300-mb winds and pressure suggested a maximum wind at that level of 180 knots (90 m sec⁻¹). At Omaha, a wind of 206 knots (106 m sec⁻¹) was reported at 227 mb.

10.5 *Life History of a Model Cyclone*

When a front is stationary, the winds blow parallel to the boundary, and there is almost no vertical motion of the air. A trough lies along the front with high pressure in both cold and warm air (see Figure 10.10a).

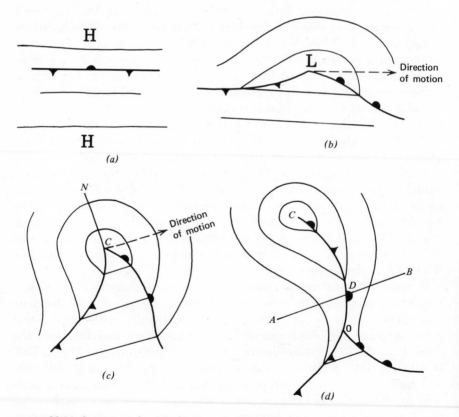

FIGURE 10.10 Stages in the life history of a model cyclone: (a) initial stage; (b) developing stage; (c) mature stage; (d) occluded stage.

Along this boundary, there is considerable energy available for release because of the juxtaposition of warm and cold air masses. This energy may be converted into other forms through a series of developments, beginning with a fall in pressure at some point along the front.

The cause for the initiation of the fall is not completely understood. It probably has some relationship with the variation in speed along the jet stream. In one area, the jet stream may be removing more air than it is bringing in, having a suction effect on the air near the ground and reducing the surface pressure. The resultant pressure gradient carries air toward the developing low. Clouds form in the rising warm moist air, releasing latent heat to the developing system. The earth's rotation transforms the flow into a cyclonic whirl, reducing the inflow and, thus, permitting the low to develop. Thus begins the second stage of the storm (Figure 10.10b).

After the cyclonic flow has developed, the conservation of rotary motion or vorticity (see Section 7.7) tends to maintain it and to transport the whirl downstream and, hence, parallel to the surface front. The motion of the center causes readjustments in the surface flow pattern. To the rear of the low the winds in the cold air push the boundary surface toward the south, forming a cold front.

In the development illustrated in Figures 10.2, 10.3, and 10.4, a low developed rapidly on the wave in Minnesota and incorporated the low of Colorado. It moved eastward, and by 1200 h GMT 9 January lay over the central Great Lakes. The arctic air pushed rapidly southward and was modified so much that the distinction between the polar continental air and the arctic air was slight. By 1200 h GMT, it covered most of the Mississippi Valley and east as far as Pittsburgh, Pennsylvania.

Normally, the cold front moves counterclockwise about the center more rapidly than the warm front (Figure 10.10c). This process may continue until the cold front catches the warm front at the surface (Figure 10.10d). There is still warm air aloft between the two frontal surfaces which slope in opposite directions; this warm air is said to be *occluded* or hidden. A trough of low pressure lies beneath the lowest point in the warm air. When this passes, there is a wind shift and, on some occasions, a change in the air-mass characteristics. This latter occurs if the cold air to the rear of the cold front has been modified somewhat differently from that ahead of the warm front. One speaks then of the *occlusion* or the *occluded front*. The energy of the storm is being dissipated by the readjustment of air masses through occlusion, and the center tends to stagnate and to fill.

In the development of 8 to 9 January, the warm front west of Lake Michigan moved slowly, but at the same time the cold arctic air advanced southward. This left a trough of warm air aloft, an occlusion, from Duluth, Minnesota to northern Illinois at 0000 h GMT 9 January.

The various stages in the life history of a model cyclone are given in Figure 10.10. There are many modifications of these stages caused by variations in the weather phenomena. Even in the sequence illustrated in Figures 10.2, 10.3, and 10.4, the low over Colorado disappeared as the northern low developed, and the two cold fronts became one. Additional heat from an underlying surface such as the Gulf Stream will influence the development, and in a mountainous area a low-pressure center can fail to keep its identity. In spite of these variations, the stages in the typical life history of many storms can be observed with a careful analysis of the data.

10.6 *Weather Associated with a Warm Front*

Ahead of an extratropical cyclone, the warm air is carried against the frontal surface. Friction and other causes keep the cold air from retreating as rapidly at the surface as the warm air advances. This tends to cause the warm air to move up the surface, and also tends to cause the slope of the front to decrease to 1/150 or 1/200. The broad stream of warm air ascends and cools to give an extensive cloud cover (see Figure 10.11) whose base is rising with distance from the front. Precipitation, as snow or rain depending on temperature, falls from the alto-stratus above the frontal surface. If the air has convective instability, the precipitation will tend to have a showery nature. Below the frontal

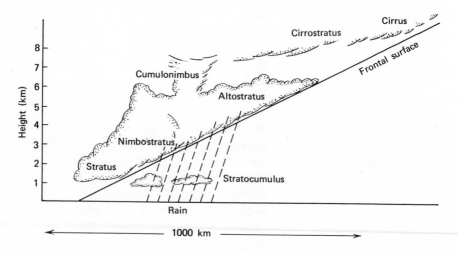

FIGURE 10.11 Cross section through a warm frontal surface.

surface, some of the rain can reevaporate into the cold air to cause it to become saturated. Here stratocumulus clouds or, close to the front, fog can form. Farther ahead of the front, 1000 km or so depending on the situation, broken cirrostratus or cirrus forms in the warm air but the condensed moisture is not sufficient to cause rain. The upslope flow over a warm frontal surface forms the family of stratus clouds (Figure 4.8), although instability may develop through lifting or radiation to produce cumuloform clouds as well.

Figure 10.9 gives information on this warm-frontal weather, giving a vertical cross section from Louisiana to Quebec at 0000 h GMT 9 January. At this time, the major warm front was at the surface near Nashville, Tennessee and sloped to the north to lie at 7100 m over Pittsburgh. The second warm front, between modified polar air and arctic air, lay south of Dayton, Ohio. Skies were clear over Quebec. Some high cloud had moved in over Ontario. South of Lake Erie the skies were overcast with middle and low cloud, and snow or freezing drizzle or rain was falling. The freezing precipitation was associated with the area where above-freezing temperatures were present above the frontal surface. The surface position of the front can be identified by a shift of wind from southeast to south and by a temperature change from the 20's north of the front to the 40's south of the front. The weather at Columbus, Ohio (see Appendix 4) illustrates the sequence of weather at a warm front. High clouds appeared at 1030 h GMT 8 January at which time the winds shifted to east-southeast. Skies became overcast with middle cloud soon thereafter, and by 1940 h GMT the base of the cloud was at 3800 ft. Freezing drizzle and rain began at 2300 h when the temperature was 23°F and continued until 0630 h GMT 9 January. At this time, with the passage of the warm front, the wind shifted to south and the temperature rose to 33°F.

The southern warm front was marked by a wind shift from southerly to southwesterly, by a temperature rise to the 50's, and a decrease in the amount of cloud.

The forecaster and students of weather maps see the total situation. The local observer has his first warning of impending change with the movement of cirrus clouds, "mare's tails," across the sky with the typical warm front. The cloud thickens to hide the sun, and rain begins with strengthening easterly winds. Normally, the rain ceases before the warm front arrives. The frontal passage itself is accompanied by a wind shift to the south or southwest, a halt to the rising temperature trend, and a breaking up of the clouds. Pressure falls during this period as the center of low pressure approaches, but this is checked when the warm air reaches the observer.

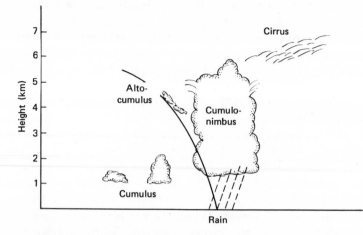

FIGURE 10.12 Cross section through a cold frontal surface.

Much detail on the weather in the vicinity of a surface front may be obtained from the weather reports made regularly on the hour and sometimes oftener at many airports. They, therefore, give a careful description of the weather as seen by the local observer. Appendix 4 contains the hourly reports for a number of stations from 1200 h GMT 8 January until 1200 h GMT 9 January. They should be studied to observe how the sequence of weather varied from station to station, both for warm fronts and cold fronts.

10.7 Weather Along a Cold Front

The effect of friction retards the cold air near the surface more than the air aloft. Thus the slope of the surface becomes steeper, approaching 1/50. The active undercutting of the warm air by the cold and the steepness of the slope of the frontal surface tend to force the warm air aloft rapidly. The presence of any conditional instability in the warm air mass augments the vertical currents to produce cumulus-type clouds. Thus, along and ahead of a cold front, there is frequently a narrow band of cumulus and cumulonimbus clouds with shower activity (see Figure 10.12).

The weather reports for 8 to 9 January provide information on the frontal activity along the cold front. At Columbus, Ohio (see Appendix 4) the polar continental air remained only from 0630 h until 0830 h, when the wind shifted to west, 14 kts with gusts to 23. Fog that had been present disappeared, and the temperature dropped rapidly from 36°F at 0900 h to 21 at 1400 h GMT. There was no precipitation at the time of the frontal passage at Columbus. The sequences of events at Indianapolis and Fort Wayne, Indiana were similar to that at Columbus. Farther south, there were light brief showers along the front when the warm air was replaced by the cold polar air. This occurred at Memphis, Tennessee at 0400 h GMT, and at Nashville at 0600 h. At Nashville the temperature dropped from 63°F to 51°F in one hour with the frontal passage. At Memphis the drop from 68°F to 52°F took two hours. The temperatures continued to drop as the cold air continued to flow over the district.

To the surface observer, there is little warning of the approach of the cold front, although he may be able to observe the band of cumulonimbus clouds on the horizon one or two hours before the front arrives. At the frontal passage the shower activity may augment the drop in temperature associated with the arrival of the colder air mass. The wind shifts to west or northwest, and pressure begins to rise as the trough associated with the cold front passes.

10.8 *Warm Air Aloft*

When the front is occluded, the observer at the ground never feels the warm air, although rain falling from the clouds in it may warm the air slightly. As the occlusion approaches, the height of the cloud base in it lowers and rain or snow often falls. Figure 10.13 gives a picture of the cross section along the line *AB* of Figure 10.10*d*. While the height of the warm air is decreasing, the wind is generally easterly.

Near the time when the warm air is lowest, the wind shifts to the westerly or northwesterly quadrant. The winds may be very light at

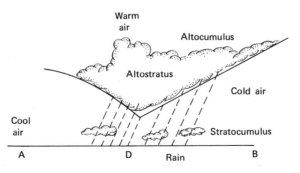

FIGURE 10.13 Cross section through an occlusion.

the trough line, and the wind shift not noticed immediately. There may be a slight temperature change if the modification that has progressed in the air to the rear of the occlusion is significantly different from that in the air ahead. For instance, when an occlusion approaches the Washington coast, the air ahead of the occlusion may have come recently from the interior and be polar continental air. To the west of the occlusion the air will have had some maritime influence and be polar maritime air. After the warm air has reached its lowest point and begins to rise, the precipitation usually becomes intermittent or ceases, but clearing is slow.

At Madison, Wisconsin the winds were east-southeast ahead of the occlusion until 0100 h 9 January, with overcast skies, snow and fog (see Appendix 4). The wind became light, and the snow grains fell until 0300 h 9 January, the time of lowest pressure. With rising pressures thereafter, the wind became west (10 to 15 kts), the precipitation stopped, and by 0800 h the sky had cleared. The temperature was 22°F at the time of lowest pressure, but dropped to 5°F at 0800 h as the cold air advanced over the station.

A similar sequence of weather can occur before occlusion begins. In Figure 10.10c the warm air reaches its lowest point along the line CN as the storm center passes to the south. The forecaster then notes on the weather map an upper front or a trough of warm air aloft rather than a surface occlusion.

10.9 Development and Movements of Pressure Centers

Figures 10.5 and 10.6 show the locations of centers of highs and lows. Except for locating the semipermanent centers of high or low pressure, there seems very little information to be obtained from these charts. Close study of many of these charts and of sequential charts gives a clue to the general movements of highs and lows.

Moving anticyclones usually are offshoots from one of the semipermanent centers discussed in Chapter 9. A cold anticyclone moving out of the polar regions generally weakens, and then merges with another high-pressure area. The movements sometimes carry them into the subtropics where they merge with the high-pressure cell there. Like low-pressure centers, they generally move with the winds at heights of 3 to 5 km.

Lows usually develop along a frontal surface. When several lows develop along the same section of a front, they are considered as belonging to one family. Although the locations in which they develop appear to be associated with variations along the jet stream, the surface conditions also enter into the development. The development of a cyclone permits a release of potential energy arising from the association of two air masses, and lows will have a tendency to develop in those areas where the instability is greatest. Often lows develop in the Atlantic off the east coast of the United States. Here the circulation introduces warm moist air off the Gulf Stream into the frontal zone, thus increasing the potential instability of the system. Figure 10.6 shows that the Gulf Stream area off North America and the corresponding Kiroshio Current area off Asia are areas of frequent storms.

Another area where lows tend to develop is in Texas. Frequently a front separates warm moist tropical maritime or equatorial air on one side and cold polar continental air on the other. The rapid development of pressure centers in November over the Great Lakes (Section 9.4) is another example of the effect of additional instability leading to intensification of a low.

When air at 3 to 5 km flows over a mountain ridge, it tends to develop a cyclonic flow on the lee side of the mountains. At times, this flow develops a surface cyclone which then moves off in the general flow. These flows are found to develop frequently in eastern Colorado and in Alberta. Initially the warm air in these lows is usually very dry and in the winter season warm, producing the chinooks that are well known to residents of these areas. As the warm air progresses across the plains it cools and picks up moisture, so that the air rising over the warm front usually becomes more moist. Lows develop in the Gulf of Genoa, partly as result of the instability arising because warm air over the Mediterranean meets polar continental air off Europe. But the flow across the Alps also contributes to their development.

The laws concerning the conservation of rotary motion influence the developments of highs and lows as they move equatorward or poleward. As stated in Section 7.7, the law is related to motion about an axis fixed in space. But the motion of the earth has its influence, as will be readily understood if we imagine a vortex located at the North Pole. The vorticity associated with the earth is added to the vorticity associated with the atmospheric flow. The vorticity that comes from the earth's rotation changes as one travels along a meridian, reducing to zero at the equator because the rotation being considered is that around a vertical axis. If, then, a cyclone moves southward, other factors remaining constant, the cyclonic vorticity around the core must increase to compensate for the decrease of the vorticity around the earth's axis. Similarly, an equatorward-moving anticyclone will tend to weaken as it approaches the equator. The opposite is true for pressure centers approaching the poles. In reality, other influences often mask this effect of meridional motion.

As stated in Section 7.8, friction tends to increase the cross-isobar flow. The effect of this flow is observed when a moving cyclone approaches a mountain area. The intensity of the low decreases over the mountainous terrain, and the active center tends to follow the passes. Thus lows from the Mississippi Valley tend to follow the St. Lawrence River or to pass eastward around the southern tip of the Appalachians. At times, the lows appear to lose their formation over the mountains only to re-form and re-intensify in the less rugged areas to the east. Few lows were observed in December 1957 (Figure 10.6) over the mountain ranges of western North America, and in this the month was typical.

Although we may discover general rules about the life histories of depressions and anticyclones, observation and study show that no two centers are truly alike. A forecaster must be constantly alert if he is to detect and to interpret the significant differences and to predict the future developments correctly.

PROBLEMS AND EXERCISES

1. When the forecasts indicate the approach of a rainstorm, take frequent (hourly if possible) observations of cloud types and amounts, wind direction and speed, temperature, and precipitation. How many of the characteristics of warm-frontal weather did you detect?

2. Many national weather services publish daily weather maps. If one is available, notice the change in weather along a line perpendicular to a warm front, and to a cold front.

3. If a series of maps is available, notice the paths of extratropical cyclones. What are their speeds? Does a storm change its rate of motion with time? Its direction of motion? Can you relate the direction to the flow of air aloft?

4. Examine the reports found in Appendix 4 and relate the reported changes in the weather to the changes on the the weather map, as shown in Figures 10.2, 10.3, and 10.4.

CHAPTER 11

HURRICANES, TORNADOES, AND OTHER TYPES OF SEVERE WEATHER

11.1 Whirlwinds and Dust Devils

The hypotheses of fronts and frontal lows, described in Chapter 10, do much to explain the cloud formations found in the extratropical cyclones. The original hypotheses have been modified with further observation and study, but they still provide a key to the cloud structures as observed by aircraft and satellites.

The instability discussed in Chapter 6 is the cause of a family of storms with characteristic cloud structures. They differ from the frontal storms in much the same manner as cumulus-type clouds differ from the stratus-type. Just as the boundary between the two cloud types is indistinct, so is the boundary between pure convective storms and extensive frontal lows.

Within the family of convective storms, the dust devil or whirlwind is the simplest, the smallest, and the most common. It does not deserve the name of a storm, but many of its characteristics are found in more intense fashion in the larger storms. A whirlwind is a small vortex in which the wind carries aloft dust, paper, and other light objects that permit the vortex to be seen. At a time when the lapse rate near the surface is super-adiabatic, a bubble of air over a relatively warm area is caused to rise by buoyancy forces. In rising, it acquires a spin, which is usually cyclonic because of the effect of the Coriolis force. Having formed, the whirlwind is in an approximate state of equilibrium, the forces present being balanced, as discussed in Section 7.7, with the core being a center of low pressure. Few measurements have been made of the drop in pressure in a dust devil, but one measurement in Arizona[1] showed a drop of 3.5 mb with the passage of a dust devil 30 ft in radius. Conservation of vorticity causes the whirlwind to move as a unit with the winds at 1 km. The vortex disintegrates when it moves over an area where the air is more stable or when turbulence in the broad air stream destroys the balance.

Some whirlwinds begin with an anticyclonic motion even though the pressure is low in the center. Under these circumstances, the balance of forces has the centrifugal force opposing both the pressure gradient force and a relatively small Coriolis force.

Whirlwinds are usually conical in shape, with a diameter 10 m or less at the base and a height under 50 m. Over hot desert areas the height may reach 100 m, and some have been observed to reach 400 m. They normally occur in regions and at times when instability is greatest, that is, over bare dry soil during the afternoon.

Although whirlwinds develop with vertical currents associated with instability, not all such currents produce whirlwinds. A light aircraft flying over heated surfaces hits "air pockets" as it passes from a region of rising air to one of subsiding air. A gliding pilot, seeking altitude, looks for thermals, that is chimneys of rising air, and in them rises smoothly for they usually show little or no evidence of rotary motion. The reasons for the development of the rotary motion are not well known. An obstruction to the flow, such as a tree or the corner of a building, can in some places initiate a circular motion that will maintain itself until destroyed by friction or turbulence. But this does not identify the initial conditions under which, over level heated ground, whirlwinds sometimes develop and sometimes do not.

[1] Sinclair, Peter C., 1965: A microbaraphone for dust devil measurements. *Journal of Applied Meteorology*, **4**, 116–121.

11.2 *Rain Showers*

When rising air in a thermal reaches its lifting condensation level, a cumulus cloud forms. As described in Section 6.6, the thickness of the cloud depends greatly on the lapse rate of the environment. Gliding pilots use the cumulus caps as guides to the location of the thermals that will carry them aloft.

As described in Section 4.8, rain falls from these cumulus clouds only under special conditions. In the moist equatorial regions, rain showers fall from cumulus clouds when the depth of the cloud has grown to 2 km or more. Under these circumstances, the temperatures within the cloud are above 0°C. In temperate latitudes, the cloud usually gives precipitation only if the top reaches the freezing level. The instability showers that occur are local in nature. An individual shower is short lived, the longest seldom lasting an hour, and moves with the wind at the cloud level. Two or more cells may combine or new cells may be initiated so that to the observer the storm appears to last longer.

Although conditional instability through surface heating or advection over warmed ground is a major cause for the formation of cumulus clouds and showers, other processes which have been discussed in the preceding chapters may initiate or intensify the currents. Near the surface, frontal or orographic lifting, horizontal convergence, or friction may contribute to the result. Aloft, divergence, as sometimes found near the jet stream, or advection of cold air can increase the intensity of a vertical current. Also, when a standing wave develops (see Section 7.13) within the atmosphere, the rising currents found in parts of the wave may augment vertical currents found underneath them. The most intense vertical currents are found when two or more causes act together. With surface heating alone, the instability develops slowly enough that vertical currents keep the situation from becoming explosive.

11.3 *Thunderstorms*

The casual observer recognizes little difference between a convective shower and a thunderstorm except that the latter is accompanied by thunder and lightning. The meteorologist recognizes that there is a marked difference in the intensity of the instability. Lightning seldom occurs if the cloud top is below 5 km, although it is not unknown. According to studies using radar, the higher the cloud top, the greater the probability of lightning.

The lightning flash is a discharge spark between a region of negative charge and one of positive charge. Studies have shown that these exist within a thundercloud with an average distribution as shown in Figure 11.1. The positive charge in the upper portion of the cloud, above the —20°C isotherm, and the negative charge below are known to exist in most shower clouds. The intense core with a high positive charge at A in the leading portion of the base is unique to the thundercloud.

Lightning discharges occur between the positively charged and negatively charged areas within the cloud or from the charged area at the base of the cloud to the ground. The former discharges occur more frequently but the latter are the strokes that attract most attention.

To explain the lightning flash, one must seek for some mechanism to separate positive charges and negative charges in the cloud. Although many hypotheses have been advanced, scientists are not agreed on their relative merits. Some see an influence of the D, E, and F layers, positively charged layers above 50 km, that is, in the mesosphere. In normal weather, they induce a negative charge at the surface of the earth and a positive potential gradient in the vertical.

The following hypotheses have been advanced to explain the physical processes by which positively charged ions can be separated from negatively charged ions and carried into different parts of the clouds. Laboratory tests have confirmed the effectiveness of some of them.

(a) Friction through collision between ice crystals leaves the air positively charged and the ice crystals negatively charged.

(b) A water drop falling toward a negatively charged plate (that is, the earth) captures more negatively charged ions than positively charged ions and becomes negatively charged.

(c) A strong vertical current will detach small droplets with a positive charge from a drop falling toward a negatively charged plate to make the drop acquire a negative change.

(d) The freezing of water droplets as they meet a large ice crystal causes a separation of charges with the large ice crystal or small hailstone becoming negatively charged while the small ice splinters ejected into the air carry positive charges.

(e) When a temperature gradient exists in an ice crystal, the warm end acquires a positive charge and the cold end a negative charge. Such temperature gradients may develop when two ice crystals of different temperatures touch momentarily, or when a hailstone which is cold in the center acquires a layer of freezing water. This may be the phenomenon that explains processes (a) and (d).

Theoretical and laboratory studies are continuing in order to determine which of these hypotheses best explain the observed phenomena of the lightning discharge, in particular, the rate at which the charge dissipated by a stroke reestablishes itself.

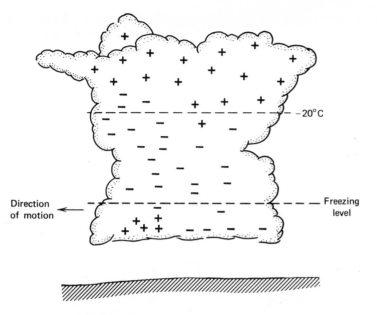

FIGURE 11.1 Distribution of electrical charges within a thundercloud.

11.4 *Distribution of Thunderstorms*

Of the processes listed in Section 11.2 leading to the release of instability, the one that leads to the development of most thunderstorms
in most areas is surface heating and the subsequent vertical current.
For certain areas other processes are more significant. In the equatorial
trough, convergence is the major cause. Along the Washington-British
Columbia coast, the orographic development of instability produces
winter thunderstorms. In the far north, the instability arising from surface heating must be augmented by another process. A thunderstorm
was recorded at Resolute, Northwest Territories (75°N 95°W), when
a cold front passed the station, and a similar combination of causes
would likely be present for most thunderstorms in polar regions. Winter
thunderstorms are reported from the oceans off the east coasts of Asia
and North America when cold air is carried off the continents at high
altitudes to increase the instability. But in comparison with the equatorial region the number of thunderstorms is relatively small.

It is estimated that at any one time more than 2000 thunderstorms
are occurring. Most develop over the land areas of the equatorial belt,
for example, over Indonesia and the basins of the Congo and the Amazon.
Some stations in Indonesia report thunderstorms on more than 300 days
per year. Away from the equator, the frequency reflects the annual

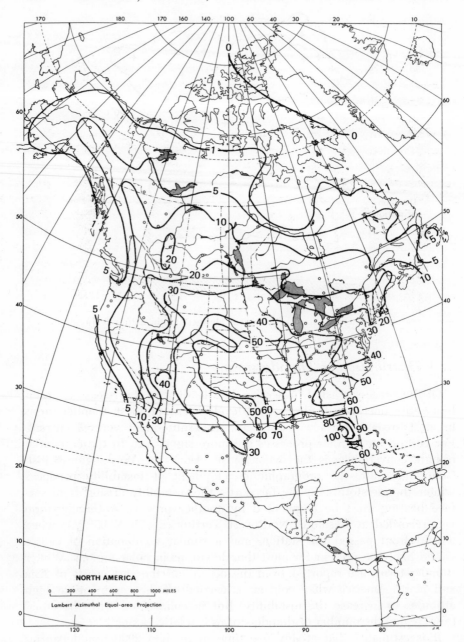

FIGURE 11.2 Mean annual number of days with thunderstorms, United States and Canada.

rhythm of solar heating. Maximum frequency is found in those areas where the summer monsoon carries equatorial air from its source to land areas: India, Malay Peninsula, Malagasy, and Florida. Desert areas have fewer thunderstorms because of lack of moisture.

Figure 11.2 shows the annual number of days with thunderstorms for United States and Canada. The maximum number is found in Florida. Along the Atlantic coast the rate of decrease is rapid, with the number dropping to less than 50 per year in North Carolina. Very few thunderstorms form along the Pacific where a cool ocean current keeps the air stable. Moving northward through the continental interior, the storms become more concentrated into the summer months. Along the Gulf Coast, about 50 per cent of the thunderstorms occur in June, July, and August. Along the Canadian boundary of the plains states, this proportion has reached 70 per cent. This change is shown in Figure 11.3. The instability, although reaching a maximum in all areas during high summer, is much more concentrated in time in the northern areas. The data for Flagstaff, Arizona show that in this desert state the required moisture and with it the thunderstorms come during July and August.

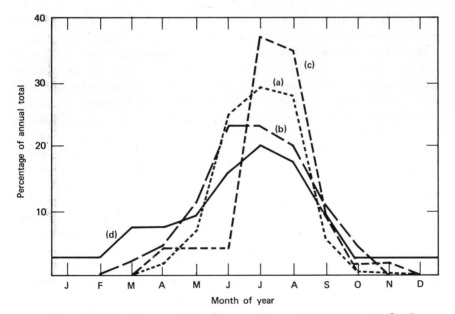

FIGURE 11.3 Annual variation of thunderstorm days at (a) Regina, Saskatchewan (total 21); (b) Huron, South Dakota (total 44); (c) Flagstaff, Arizona (total 46); (d) Pensacola, Florida (total 69).

11.5 *Hailstorms*

Hail comes from tall cumulonimbus clouds. The higher the tops of the clouds, the greater the probability that hail will form. In Alberta, Canada the probability of hail is more than 80 per cent if the cloud top reaches 12 km. Farther south, the cloud tops of hail clouds are generally higher.

The special feature of a hail cloud is the formation of large ice spheres (Figure 11.4) which fall to the earth. When the temperature is low and the amount of moisture is small, the hail is less than 1 cm in diameter (shot or pea size). Although small, a heavy fall can cause considerable damage to crops, especially when accompanied by a high wind.

Figure 4.15 shows the picture of a cumulonimbus cloud near Penhold, Alberta. The instability that gave rise to the cloud is shown by the plot on a tephigram of the radiosonde ascent in Figure 6.8. The storm gave hail over an area 40 mi long with a width from 1 to 10 mi. The hail area was delineated by a careful survey of the farmers living in the district. Figure 11.5 shows the area on which hail fell. It also gives the outline of the radar echo from the storm at 1850, 1909, 1929, 1949, 2010, and 2030 h 15 July 1968. Notice that the storm moved eastward during the period of observations. Also the area on which hail fell lay along the southern boundary of the echo. This is the usual situation, although hail does occur in other sections of the storm track.

Hailstones usually consist of concentric spheres of alternately clear and opaque ice. As described in Section 4.9, an ice crystal grows through sublimation, or by collision with water droplets. Supercooled water droplets tend to freeze on collision, with the latent heat passing to the surrounding air. The ice structure formed depends on the rate of deposition. If the rate is small, the water droplets will freeze separately and the resultant ice will contain some entrapped air bubbles, forming an opaque ice. If the deposition is rapid, or if the water collects at temperatures above freezing with the stone moving to colder regions later, the new deposit will surround the initial sphere and will freeze as solid clear ice. The turbulence within a large cumulonimbus cloud is sufficiently great to carry growing stones alternately into the strong updrafts where the liquid water content is great and then into other regions where the water droplets are less plentiful and smaller.

To contain the necessary moisture and to develop vertical currents strong enough to sustain the growing hailstones within the cloud against the pull of gravity, the air must be moist and instability great. The vertical winds must be of the order of magnitude of 100 m sec^{-1} or 200 mi hr^{-1} if the largest stones are kept aloft when they are falling

FIGURE 11.4 Hailstones. (Published by permission of R. H. Douglas.)

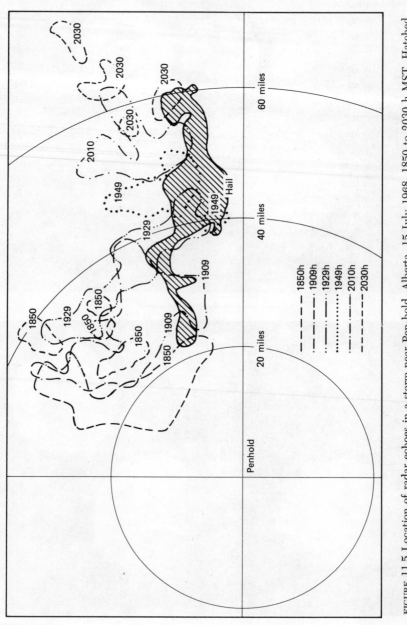

FIGURE 11.5 Location of radar echoes in a storm near Pen-hold, Alberta, 15 July 1968, 1850 to 2030 h MST. Hatched portion gives the area on which hail fell during the storm.

under the influence of gravity. The instability necessary to produce these currents can and does at times occur in the atmosphere. If the storm is stationary so that the stone falls through the cloud, the warm cloud droplets below the freezing level tend to melt the stone. Hail is therefore not common under these circumstances. Usually, it seems, the hailstone is carried forward out of the cloud top and there, away from the vertical currents, falls to the earth. Some stones are caught in a rising current near the cloud base and enter the cloud again to grow larger.

Another contributing cause to the development of hail is the forced lift at a frontal surface, particularly that of a cold front. The horizontal convergence in the vicinity of a low-pressure area adds to the vertical currents and increases the risk of hail. Another cause has been suggested, arising out of the geographical distribution of hail. Although hailstorms occur in many areas from the equator through the region of the westerlies, maximum severity is often found in the lee of high mountain ranges. Among these regions are the Po Valley of Italy, along the east coast of New Zealand, eastern Colorado, southern Alberta, and western Argentina. It is quite possible that the frequency of hail in the lee of mountain ranges is associated with the vertical currents found in the standing waves in these regions (see Section 7.13).

A strong wind at high levels, which may be part of the jet stream, seems to contribute at times to instability and vertical motion. Such wind effects were incorporated into the model of a hailstorm proposed by K. A. Browning and F. H. Ludlum (Figure 11.6), a model which explains many of the observed phenomena of these storms. The strong wind carries growing hailstones ahead of the main cloud where they begin to fall to the earth. The lower levels of the cloud catch the falling stones and another updraft carries them aloft, meanwhile adding to the mass of the stones. The cycle may be repeated, but finally the weight of the stones or the absence of an updraft permits the hail to fall to the earth. Such a model explains why, on most occasions, the hail is the first precipitation to fall from the cloud. Other scientists who have studied hail believe that there are other processes by which hail develops and grows. As yet, no one really knows the answer.

11.6 Tornadoes

The most violent of weather phenomena is the tornado (Figure 11.7). It does not cause the widespread damage of a hurricane or an ice storm; but in the path of a tornado the destruction is complete. It, too, forms through vertical currents that are caused by instability. As with hailstorms, thermal instability is usually not sufficient, since this tends to

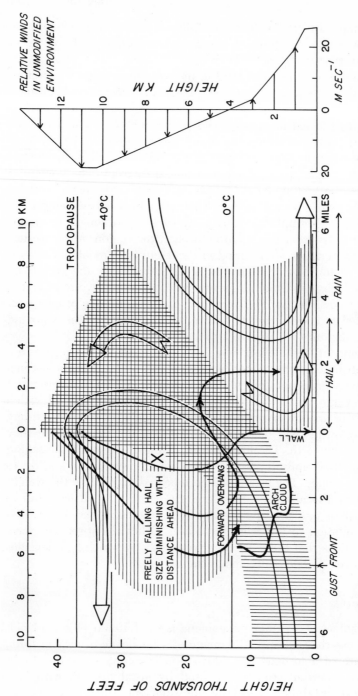
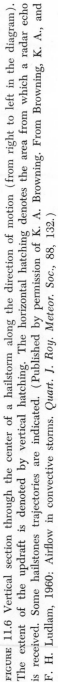

FIGURE 11.6 Vertical section through the center of a hailstorm along the direction of motion (from right to left in the diagram). The extent of the updraft is denoted by vertical hatching. The horizontal hatching denotes the area from which a radar echo is received. Some hailstones trajectories are indicated. (Published by permission of K. A. Browning. From Browning, K. A., and F. H. Ludlam, 1960: Airflow in convective storms. *Quart. J. Roy. Meteor. Soc.*, 88, 132.)

be released before it reaches the point where it would initiate a tornado. The additional factor is usually the vertical motion along an active cold front approaching the district. The jet stream aloft associated with the front may also contribute to the development of a vertical current.

In Section 6.6, stability was discussed by using the radiosonde ascent over Oklahoma City at 1800 h CST 5 June 1962. Figure 11.8 gives the weather map for 0600 h CST of the same date, a day on which a number of tornadoes hit the state of Oklahoma. From this map would be prepared the forecast for tornadoes later in the day. Notice the front lying over western Texas and Oklahoma. To the west was cool, dry air as indicated by the dew points plotted at the lower left of the stations. The air to the east was both warm and moist. Thunder was reported during the night (see lower right of station circle for past weather) at a number of stations, giving evidence that the air was unstable. This would be confirmed by a study of upper-air temperatures. The vertical

FIGURE 11.7 A cumulonimbus cloud with the funnel of tornado. [Courtesy, Eric Lantz, Walnut Grove (Minn.) Tribune.]

currents arising from thermal instability would be increased because of the approach of the cold front and because of strong winds aloft.

The strong vertical currents in a cumulonimbus cloud, reaching at times 25 m sec⁻¹, can develop a vortex similar to the vortex seen at the outlet of a bathtub. In some clouds, this vortex remains hidden. In others, the vortex is seen to protrude from the base and descend to the ground. Occasionally a cloud will have more than one vortex (or funnel) below the main cloud deck. The winds in these funnels have not been measured but are estimated to reach 100 to 200 m sec⁻¹ (200 to 400 mi hr⁻¹). They are caused by an extreme pressure gradient toward the low pressure at the center of the vortex. The pressure at the center is unknown, but on 24 May 1962 a barometer located one mile from the path of a storm fell 34 mb as the tornado passed. It is probable that the central pressure in moderate or severe tornadoes is below 800 mb.

When the funnel touches the ground, the combined effect of the rapid

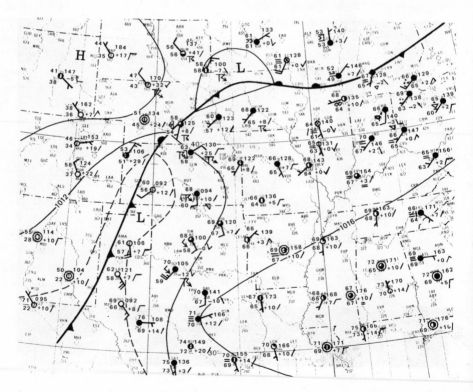

FIGURE 11.8 Weather map of 0600 h CST 3 June 1962, a day when ten tornadoes hit the state of Oklahoma.

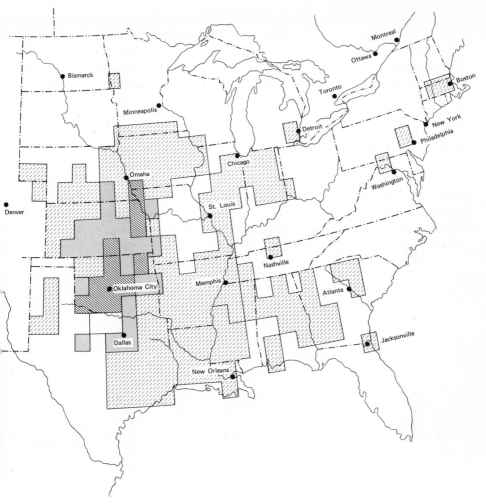

FIGURE 11.9 Distribution of tornadoes in the United States in mean number per decade per degree square, 1916 to 1961 (after Roth) (light shading 5–10; moderate, 10–15; heavy, 15–20.) (From Roth, Richard J., 1963: Severe local weather and the property insurance industry. *Amer. Meteor. Soc., Proc. Third Conf. on Severe Local Storms.*)

and extreme drop in pressure and the high winds destroys any obstruction in its path. The vertical currents can lift roofs or even houses and deposit them 50 m or more away from their original sites. The average width of the path of destruction is about 400 m, and the length 20 km, but there are wide variations. A few tornadoes have been observed to rise from the ground and then touch down again.

Tornadoes occur in regions where instability is great and an active cold front can give extra impetus to the vertical currents. The Indian subcontinent and Australia are both regions of frequent tornadoes. But the Mississippi Valley, where moist air off the Gulf of Mexico can meet polar continental air from the Canadian prairies is the region where they are most common. Figure 11.9 gives the frequency, per 1° square in 10 years, and shows that the core of maximum frequency lies in Oklahoma and Kansas. The annual trend of frequency (see Figure 11.10) is related to the southward extent of extremely cold air. Along the Gulf Coast, tornadoes occur in all months, with a slight maximum in March and April. Farther north, in the region of maximum frequency, the worst time of year is May, with very little risk after June. In the block of states along the Canadian boundary, June is the worst month, and the occasional tornado north of the border usually occurs in July.

Water spouts are similar in formation to a tornado, but with the funnel

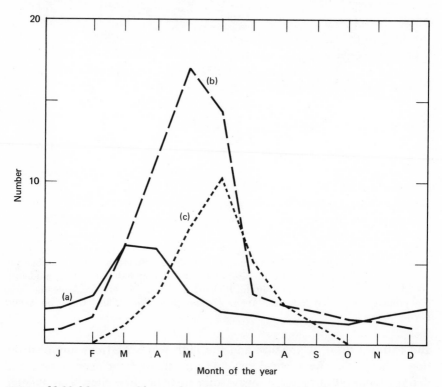

FIGURE 11.10 Mean monthly number of tornadoes: (a) Florida, Georgia, Alabama, Mississippi, Louisiana; (b) Oklahoma, Arkansas, Kansas, Missouri; (c) Nebraska, Iowa, South Dakota, Minnesota.

forming over the ocean or occasionally over an inland lake. They are relatively weak and of short duration, averaging 15 minutes. They can damage ships, but the extent is minor compared to that from a tornado. They too emerge from the base of cumulonimbus clouds, frequently clouds that have been swept off an adjacent land area. They are common in tropical waters, such as the Caribbean and the Mediterranean.

Tornadoes and hailstorms are formed during the release of potential energy that is contained partially in the unstable situation resulting from solar heating, and partially in the latent heat of the water vapor present in the warm surface air. Part of the released energy is transformed temporarily into the kinetic energy of high winds, but this gradually dissipates itself into sensible heat. The process does little in transferring energy poleward, but it does move large amounts of energy from near the earth's surface to the upper levels of the troposphere. One may learn the order of magnitude of this release by a study of the rain that falls. One cell of a hailstorm that hit greater Miami, Florida on 29 March 1963 gave approximately 2×10^{16} cm^3 rain. The release of energy from the latent heat of condensation (see Exercise 3, Chapter 1) is approximately equal to that from a 1000-megaton bomb. The release of energy of a tornado must be considerably higher, although not so great as that from a hurricane.

11.7 *The Intertropical Convergence Zone*

The geostrophic wind equation, Equation 4, Section 7.6, gives a relationship between the wind speed and the pressure gradient force. Near the equator the term sin ϕ of this equation becomes very small and the relationship cannot be assumed to hold true. Because of this, the isobars no longer provide information on the wind flow within about 10° of the equator. Lacking the help of the isobars, the meteorologist relies more on the wind observations as guides when drawing stream lines in the tropical regions.

A common feature of the weather map for the tropics is a trough of lower pressure found along the line where southeast and northeast trades meet. This zone, the *intertropical convergence zone*, is a region of rising air currents, the driving force being partly the low-level convergence and partly the conditional instability present in the equatorial air.

The mean positions of the zone for January and July may be estimated from the maps, Figures 3.10 and 3.11, giving the mean pressures. There is a tendency for it to follow the sun northward and southward, but in the open ocean of the Pacific this tendency is slight and the zone seldom is more than 5° away from the equator. In eastern Africa, where land straddles the equator, the annual cycle carries it 10 to 15° on either side. The summer lows of northern India and Australia cause the zone to move farthest from the equator. In these areas the trade winds change their direction as they cross the equator because of the change in the Coriolis force. The southeast trades of the southern Indian Ocean approach India from the southwest, and in northern Australia the summer trades are northwesterly.

Convergence along the equatorial trough is a maximum in the region of *easterly waves*. They are found in regions where the convergence zone forms a westward-moving wave with its crest on the poleward side. They are areas where the clouds are thickest and the rain more frequent and more intense. In Hawaii, 75 per cent of the rain comes in well marked periods of rainy weather, that is, during periods when easterly waves are passing.

Aloft, the two subtropical high-pressure belts lean toward the heat equator and merge above 8 km, causing the winds above the trade-wind belts to be light westerly in relatively warm air. This stable layer suppresses the vertical development of the cumulus clouds in the trade-wind belt, with the result that clouds decrease in thickness as one moves farther from the intertropical convergence zone.

11.8 *Tropical Storms*

The disturbed weather of an easterly wave may persist for several days. If, in its movement, it should move about 15° away from the equator, the Coriolis force diverts the converging winds, causing them to take more circular paths. In this process we have the birth of a tropical storm. Not all easterly waves develop in this manner, but it is possible to trace most of the storms that develop back to the period when they were easterly waves. Some West Indian storms have been traced across the ocean to waves that have moved off the Sahara, but the energy that produces the high winds of these storms comes from the moisture that has evaporated from the warm waters of the equatorial Atlantic.

A hurricane of the 1964 season called Dora was traced back to a period of unsettled weather in the Cape Verde Islands on August 28. This area of convergence moved westward, and became a well defined storm on 1 September at 12°N 47°W. Figure 11.11 shows a picture of the whirling mass of clouds at 22°N 60°W on 4 September, taken from a satellite. The central pressure on 1 September was 998 mb, but

FIGURE 11.11 Hurricane Dora as seen from a satellite, 4 September 1964. (Tiros VIII Photograph. Courtesy U.S. Weather Bureau.)

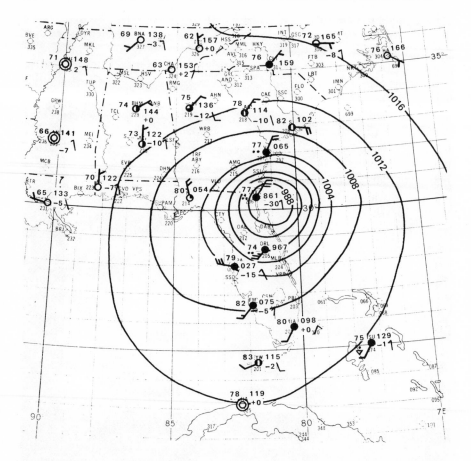

FIGURE 11.12 Weather map for 0100 h EST 10 September 1964 showing hurricane Dora off Florida Coast.

with the movement west-northwest it decreased to 942 mb on 6 September. It continued its course, to reach the Jacksonville area of Florida on 10 September (see Figure 11.12), but weakened slightly. The winds dropped from 150 mi hr^{-1} (65 m sec^{-1}) on 6 September to 125 mi hr^{-1} (55 m sec^{-1}) as it hit the Florida coast. Even with this decrease of wind speed, it inflicted damage estimated at 250 million dollars. Thereafter, the storm center changed its course to travel northeast, passing over Cape Hatteras on 13 September and Newfoundland on 15 September. The movement over Florida and Georgia weakened the pressure gradient, with the central pressure of the storm rising to 991 mb at 0100 h EST 11 September and to 999 mb 24 hours later. The moisture was still present aloft and caused excessive rains over the Carolinas.

The major events of the life history of Dora are duplicated in many of the hurricanes of the world. Beginning as an area of disturbed weather in the trade-wind belt, the storm moves far enough from the equator to develop a cyclonic flow. Moving westward, it intensifies, frequently developing winds of hurricane force. Caught in the flow around the large oceanic subtropical anticyclones, the storm moves northward or northwestward along the western boundary of the ocean. This trajectory often brings the storm close to or over land areas where it causes great damage. Passing over land areas, the hurricane loses some of its destructive force. The loss is more rapid over mountainous areas like Cuba than over flat moist areas such as the Mississippi Valley. If it moves again over warm tropical waters, it picks up more moisture from which it derives new energy. If the track carries it over cooler water, this causes the storm to lose its intensity, although it may continue as an extratropical low.

The winds around the center of the storm spiral inward and are carried upward (see Figure 4.17), releasing vast amounts of water. The circulation fails to penetrate the exact center of the storm (Figure 11.13).

FIGURE 11.13 Wall cloud of Hurricane Esther 16 August 1961. (Photo from National Hurricane Research Laboratory, Miami, Florida. Published by permission.)

Here, in a circle 10 to 40 km in diameter, the winds are light, the clouds are scattered to broken, and the temperature is high. The changes from high winds and rain to calm with broken clouds and again to high winds, in a short period, are very dramatic to one located in the path of the *eye*. The period of calm may last 30 minutes before the winds start to blow from the opposite direction.

11.9 *Regions of Occurrence of Hurricanes*

As noted in the preceding section, hurricanes are associated with the poleward flow around the western ends of the oceanic high-pressure belts. Thus they are found in the West Indies and vicinity. The winter season is not completely free of hurricanes, but few occur before July. For the 14 years 1954 to 1967, the average number of hurricane days per month for the Atlantic were as follows:

June	1	October	$7\frac{1}{2}$
July	$1\frac{1}{2}$	November	$\frac{1}{2}$
August	6	December	$\frac{1}{2}$
September	17	January	$\frac{1}{2}$
Total			34

The number of distinct storms averages 10 per year. There is a tendency to shift locations, with the early and late storms occurring in the Gulf of Mexico, and the September storms staying in the Atlantic east of the West Indies. A few storms associated with the Bermuda high are found along the west coast of Mexico moving northwestward. Over the cooler waters, they seldom develop into destructive storms. Figure 11.14 shows the paths of the tropical storms of the western Atlantic during 1964.

The Philippine area of the Pacific is even more susceptible than the Atlantic to tropical hurricanes. These *typhoons*, as they are called, are a threat to the Philippines, to the east coast of China, and to the offshore islands of Taiwan and Japan. They occur with almost equal frequency in the months of July, August, September, and October, on the average three or four tropical storms per month. The tropical storms in the Bay of Bengal are distributed in about the same manner, but the number is reduced to one or two a month. Farther west, the Arabian Sea storms, like the Gulf of Mexico storms, have peaks of frequency in June and October with very few during the height of the monsoon. A few hurricanes move northward in the central Pacific, in the vicinity of Hawaii, when the Pacific anticyclone is split into two cells.

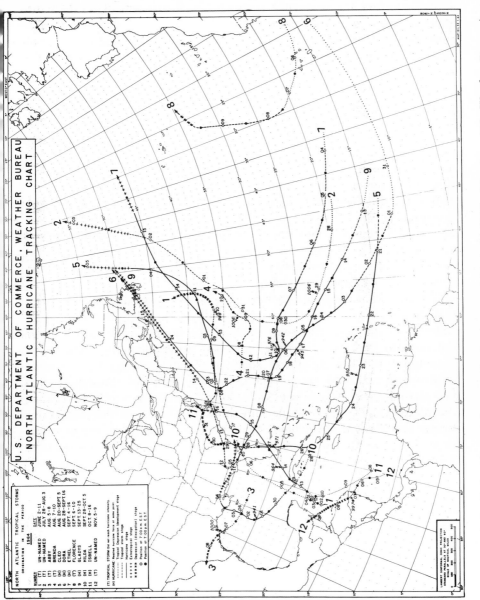

FIGURE 11.14 Paths of tropical storms in the western Atlantic in 1964. (Courtesy, U.S. Weather Bureau.)

255

Hurricanes develop south of the equator, associated with the Pacific anticyclone, but the frequency is much less than north of the equator. Some of them pass southward along the east coast of Australia, but more of them pass along the east coast of Africa. Hurricanes are unknown along the east coast of South America, possibly a result of the shape of its coastline.

11.10 *Energy of Hurricanes*

An appreciation of the magnitude of the energy conversion in hurricanes can be obtained from the amount of rain that falls. Synoptic and theoretical studies have shown that the rain from a moderate hurricane is of the order of magnitude of 10^{11} gm sec^{-1}. This represents a release of 6×10^{13} cal sec^{-1}. The energy released by a megaton nuclear bomb is approximately 1.2×10^{15} cal. Thus the energy released by the condensing moisture in a hurricane is equivalent to three megaton bombs per minute. The damage from the hurricane is less dramatic because the time and space involved are both much greater than the ones involved in the explosion of a bomb.

The energy conversion in a hurricane is, of course, the same process that goes on in the equatorial zone at all times. Solar heat evaporates water from the land and sea. Vertical currents carry this aloft where the released latent heat warms the upper atmosphere. In a hurricane, because of the Coriolis force, some of the released energy becomes transformed first into kinetic energy which dissipates through friction into sensible heat. Wind speeds of 160 mi hr^{-1} (70 m sec^{-1}) have been recorded by anemometers specially built to withstand the destructive forces. Calculations have shown that the kinetic energy at any one time in a moderate hurricane is approximately 10^{23} ergs, equivalent to 2.4×10^{15} cal. The energy being released is so great that it could quickly redevelop the winds if they were suddenly to drop to zero. Experimental attempts and theoretical studies have been made to discover if cloud seeding would distribute the energy over a wider area and so reduce the damage; but as yet no clear-cut conclusions have been reached.

11.11 *Spells of Weather*

The average weather, or the *climate*, varies from one part of the world to another. The basic laws for determining the climate are the same as the ones that determine the weather. Some of the geographical variations that influence these basic laws have been discussed in the preceding chapters. Man has learned how to adapt himself to the variations in weather and climate and, hence, to live on most of the land areas of the globe. Even so, certain weather phenomena can do considerable damage to his plans. Among them are a late spring frost, a hurricane, a hailstorm, a tornado, a sudden heavy rain storm, an early autumn snowfall, and a mild spell in late winter to start plant growth prematurely.

Other weather sequences can be less dramatic, but cause equal disruptions and hardship to mankind. A few fine days come in the normal cycle of fine and stormy weather, but sometimes the cycle seems to break down. Day after day passes without rain. The sun in a cloudless sky warms the ground excessively and evaporates the moisture. Plants wilt or just manage to stay alive without growth. Winds pick up the dry ground and carry the soil away in dust storms. Over a wide area the crop is a failure. Irrigation to combat lack of rain began in Egypt in the time of the Pharaohs. During the past 100 years, there has been a multitude of irrigation projects, but the extent to which irrigation can be used is limited.

The severity of a drought must be assessed in terms of the normal rainfall because this determines the crops that are grown. A 20 per cent drop from normal rainfall in the Malay peninsula leaves it with 200 cm (80 in.) of rain. Yet this drop will damage the rice crop of the area, just as a 20 per cent drop from the normal precipitation of 50 cm (20 in.) in western Kansas damages the wheat crop grown there.

The immediate cause for most periods of abnormal weather is the flow in the upper troposphere. Figure 3.9 shows a map of the flow at 500 mb (approximately 6 km) at 1200 h GMT 8 January 1969. The map shows that the flow at that time had marked meridional components with a low near Churchill, Manitoba, southeasterly winds northeast of Hudson Bay, and very strong north to northwesterly winds over Alaska and the eastern Pacific.

These flow patterns change from day to day. Also, their position at any time controls somewhat the movements of lows and highs. Although the flow at upper levels does change, there are preferred positions. In winter, there is usually a trough between 60°W and 80°W, with a ridge at 110–130°W. This flow, and its influence on the surface weather patterns, causes the mean isotherms (see Figure 2.11) to dip southward in eastern North America. Also, the flow carries the storms along the Atlantic seaboard toward Greenland, depositing rain or snow along the coastal area. The normal surface anticyclone of western North America is associated with the pressure ridge at high levels.

Although there are preferred positions for the upper-level troughs and ridges, for a period of weeks they can be found in other places, or be less or more prominent. In November 1963 (see Figures 9.1 and 9.2), the ridge over the continent was very weak, and the month saw a period of warm weather over the United States. The next month, December (Figures 9.4 and 9.5), had a very pronounced ridge in western North America, and gave a period of extremely cold weather. If the abnormal flow causes a series of storms to pass through an area, more precipitation than usual will fall. Oregon and Washington experienced such a situation in December 1964, resulting in serious flood conditions. When an abnormal upper-level flow keeps a ridge over the region, storm centers will be diverted from the area, the sun will shine day after day from a cloudless sky, and rainfall will be light or zero until a change occurs in the overall flow.

Meteorologists are able to recognize the influence of the upper-level flow on the weather pattern. They are not agreed as yet on the reasons for the flow to follow a particular pattern. The heat balance of the earth plays its part. If the large water area of the Pacific becomes warmer in comparison to the land, more intense lows will form and greater flows of more moist air will move over the Pacific coastline. The flow over the Pacific sets up a sinusoidal pattern so that a ridge lies over eastern North America. The resultant southerly flow brings mild moist air over the Mississippi Valley, and hence a period wetter than normal. When the Pacific waters are relatively cold, the resultant ridge over western North America causes northwest winds in the lee of the mountains, resulting in subnormal rainfall over the plains states. Once again we observe that the heat balance of the earth affects the weather patterns. But the interrelationships are many and complicated. Meteorologists are a long way from understanding them, and until they do there is little prospect of making accurate long-range weather predictions.

CHAPTER 12

WEATHER AND MAN

When we consider the effect of weather on man's various activities, we must distinguish between weather and "average weather" or climate. The climate of a country is the weather we normally expect, with its changes from month to month throughout the year. Man adjusts his activities to the climate in which he lives. A farmer in an area with an average frost-free period of 90 days grows crops that can stand a light frost. An engineer assures himself that the rainfall is sufficiently plentiful and reliable before he plans a dam across a river to provide power. A victim of arthritis moves to a dry hot climate to relieve the suffering. An airline company plans its schedules based on the average winds between the stops.

After man has adjusted to the climate, he then must struggle with the variations (the day-to-day weather) that influence his activities. In some places and in some fields of endeavor, the effects are small, but in others the variations in weather can bring disaster.

12.1 *Weather and Agriculture*

The farmer, more perhaps than any other, is affected by the weather. He plans his activity for the day or for the next few days according to the weather he expects. He has learned from experience to read

259

the weather signs so that he can make 12- to 24-hour forecasts that are reliable most of the time. Also he notes signs that cause him to amend the official forecast when this is going wrong. He could use with profit a reliable forecast for the coming month or season if it were available.

The rhythm of fine and stormy days that is found in the region of the westerlies sets the pattern for many agricultural activities. In the spring some of the land cannot be cultivated while it is wet. Yet the farmer must plant his seed early to use the spring rains, the winter moisture and, for some of his crops, the total growing season to bring them to maturity. Some operations (for example, harvesting) demand fine weather for successful completion. Farmers in irrigation districts are on the alert for possible rain which, for some crops, is cheaper and more beneficial than an equal amount of water supplied by irrigation, but can damage others severely.

The agriculturalist has learned to meet the problem of unseasonal frosts. In an area where frosts limit the growing season, he adjusts his crops to make use of the total growing season expected. Plants are started in greenhouses, to be transplanted to the fields, "when all danger of frost is past." In practice this phrase is misleading. The gardener cannot wait that long, but must be ready to plant when the probability of frost has dropped to a small value. He gambles that the frost damage in the occasional cold year will be compensated for by better crops with the longer growing seasons of other years.

In some regions the gardener or orchardist has learned to protect himself against unseasonal frosts or, in other words, he has changed the climate by increasing the frost-free period. As described in Section 5.6 and elsewhere, the nocturnal cooling is greatest at the ground level, and this cold air tends to flow to the bottom of hills. Both influences tend to develop an inversion in the air near the earth's surface. In the late spring and early fall, the top of the inversion is frequently above freezing while the ground is below freezing. Orchardists and gardeners can then protect their crops by creating currents that mix the air aloft with the lower layers. This has been done by means of aircraft propellers mounted and run so that the surface air is forced upward. A more common method is to induce convection currents with flare pots in the orchards (see Figure 12.1). They provide some heat to the air but are also effective in producing down-currents from the warm layer aloft. These methods of frost protection are beneficial in some situations but not in all. When a cold outbreak brings air with subfreezing temperatures at all levels over a district, this air passing over warm ground is unstable. Any heat supplied is carried aloft rapidly, and turbulent currents will

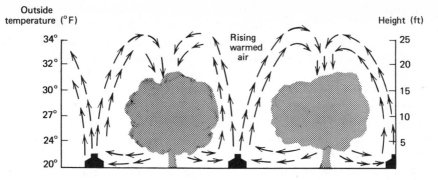

FIGURE 12.1 Currents developed by heaters within an orchard to protect it from frost.

bring cool air down to the ground. The foregoing methods of frost protection will not be useful in this situation.

In some areas of marginal rainfall, farmers and especially stock ranchers have sought help from people who claim to make rain. An inch or two of additional rain at a critical time can increase the return markedly. With this possibility in mind, some groups of farmers and ranchers gambled by engaging meteorologists to seed clouds, although research scientists were not yet satisfied that cloud seeding could produce additional rain under certain circumstances. Other farmers in regions where hail is a hazard have put money into operations designed to suppress hail by cloud seeding. Because one severe hailstorm can cause millions of dollars damage, the success of the operations need not be high for the operation to be economically successful. Still unanswered is the question of whether the cloud seeding might make a hail storm more severe under certain circumstances.

The rancher is less sensitive to the weather than the farmer, but still finds occasion when he must adjust his operations because of the elements. In areas where stock is left outside during the winter, a severe blizzard or deep snow will threaten the livestock. The rancher may then find it necessary to provide feed until the situation has eased. A cold rainy period or an unseasonal snow after the young are born can cause extensive losses unless the rancher can provide some shelter for the newborn animals. In the search for more pasture, ranchers have moved into areas of marginal and variable rainfall. To make this land more productive he, too, has called on the commercial cloud seeders to help augment the rain that is, at times, desperately needed.

12.2 *Weather and Aviation*

Through the centuries, mankind learned gradually some of the causes of the weather and weather change. The invention and development of airplanes accelerated the development of meteorology rapidly, partly because they provided additional weather data, and partly because they required more accurate forecasts. With the large financial investment in aircraft, the demands for more reliable forecasts have forced governments to spend more money on weather services than they otherwise would have done.

Previous to the invention of the airplane, balloons carried men and instruments into the atmosphere, and some moderate flights were made with them. These flights provided information on the temperature and wind distribution above the earth's surface. The information was spasmodic and governed by the weather itself because it was impossible to launch balloons in all kinds of weather conditions.

The first airplanes could not be flown in all kinds of weather, but the demands of scheduled flights resulted in a more complete knowledge of the three-dimensional picture of conditions in the troposphere and later in the stratosphere. The pilot also asked the weatherman for a forecast of temperature and winds along his route, which in turn led to the rapid development of the radiosonde network. The information gathered by the radiosondes has been useful in many of the advances in the science of meteorology during the past twenty years. Meanwhile, the advances in aircraft design and in safety devices for aircraft travel have reduced considerably the dependence of the pilot on the meteorologist. Yet it is still necessary for pilots of smaller aircraft to be assured that the weather conditions will permit them to operate in safety.

The primary need of the pilot at the present time is a forecast of the weather at his destination. Surface wind speed and direction, type of precipitation if any, and especially visibility are all important because under certain conditions the aircraft is unable to land. This demand has forced the meteorologist to study in more detail the variations in these phenomena and the causes for them, but he still has unsolved problems. The impossibility of perfect forecasts and the need of the pilot to be able to land at almost any time have forced aircraft companies to develop radar and other instruments by the help of which airplanes may be brought to earth under very unfavorable conditions. Small planes are usually not equipped to land by instrument procedures and must be assured of good ceiling and visibility conditions at the time of landing.

Flights of airplanes into clouds showed that the cloud particles were often supercooled water droplets, a result that was quite unexpected.

These water droplets froze on the airplane's surfaces, particularly on the leading edge of the wing. The change in the shape of the wing reduced the effectiveness of the plane, and a number of accidents resulted from aircraft icing. This danger was met by a careful study of the incidence of icing and the regions where it was probable. Forecasters warned pilots to avoid these areas if possible. But the study also taught the meteorologist much about the microphysics of clouds. Other research has provided the pilot with means of preventing the formation of ice on the surface of the aircraft and of breaking it off when it does form. Pilots have lost much of their fear of icing except when they must land in freezing rain.

Aircraft pilots continue to be interested in the winds at the different levels of the atmosphere even though the increased speed of the commercial aircraft has meant that these upper winds are not so vital as formerly. On the other hand, the rapid consumption of fuel gives the pilot less time to maneuver before he must land. An accurate forecast of winds at different levels permits the navigator to choose the level most suitable for the flight. Even with the high speeds of the modern jets, it is beneficial to make use of favorable tail winds and to avoid heading into the high winds of a jet stream.

A new problem for air travel has developed with the advent of high-speed jet aircraft. In certain parts of the stratosphere, pilots find themselves in areas where the turbulence can disturb the comfort of the passengers, and at times even endanger the plane itself. This *clear air turbulence,* as it is called, is under study but as yet no sure method has been discovered to warn the pilot where it may be found and how it might best be avoided. The jet stream is apparently the cause of some, but not all, clear air turbulence.

12.3 *Weather and Forestry*

In general, the activities of foresters are not greatly influenced by the day-to-day changes in the weather. But forest fires are a major cause of disaster in forests, and the meteorologist can provide assistance in fire prevention and fire fighting.

Ever on the alert to the danger from fires, the forester has studied the dampness of the forest duff (the surface layers of leaves, needles, twigs, etc., on the forest floor) and the causes for change in moisture. When the duff is wet, accidental fires from campfires or lightning die out. When it is dry, a slight spark can start a fire that will travel rapidly. The dampness of the duff is related partly to the forest conditions, but the changes are also closely related to three weather elements:

temperature, humidity, and wind speed (see Section 4.11) which determine the rate of evaporation. If this evaporation reduces the duff moisture, the fire hazard increases. Of course, rain will add moisture to the forest floor, and so reduce the risk of fire.

Foresters have produced tables that relate the changes in the dampness of the duff to these three weather elements. The rangers measure the weather elements and estimate from the tables the change in the duff moisture and, with this, the fire hazard of the forest area. Although the procedures are similar for all forest areas, fire-hazard tables vary from region to region depending on the type of forest, the latitude, and other factors. The fire-hazard index alerts the forest ranger to the risk that fire will move rapidly and causes him to prepare for a serious fire when it is high. When he considers it necessary, he collects equipment, alerts men to be prepared for fire fighting, and issues a warning to the public against open fires. In extreme situations, he closes the area to the general public.

Many forest fires are started by lightning. Dry lightning, or lightning which is not accompanied by rain, is very serious when the fire hazard is high. Warning of the possibility of lightning comes from the meteorologist who assesses the probability from the stability forecast for the day. This probability is determined partly by a study of the upper-air tem- of the weather situation that cause decreasing stability. The forester peratures (see Chapter 6) and partly by studying the other features can adjust his preparations for a possible fire according to the forecaster's assessment of the risk of lightning. In some areas, foresters have used cloud seeding to reduce the frequency of dry lightning.

After a fire has started, its spread is determined largely by the wind velocity. Protection by back firing and evacuation of personnel, if necessary, can be done effectively if an accurate forecast of wind and wind shifts is available. Although such a forecast would prove very useful, it is difficult to obtain. Forest areas are usually mountainous areas. Both the local influence of valleys and hills, and the variation of wind with altitude, make it difficult to have an accurate knowledge of the wind. To provide useful forecasts, the forecaster must make a careful study of the topography and weather characteristics of the area that he is supervising.

12.4 Weather and Hydrology

The science of *hydrology* deals with the distribution of water over the face of the earth. Meteorology, dealing with the water in the atmosphere, meets the science of hydrology at the points of transfer, that is, when rain or snow falls from the clouds above and when evaporation carries moisture from the earth's surface into the atmosphere. It is obvious that some meteorological phenomena are related to the problems of the hydrologist. The study of these phenomena and the resulting problems is called *hydrometeorology.*

Water passes through different phases of the *hydrologic cycle* (Figure 12.2), which describes the different processes by which water on the earth goes from one phase to another or from one environment to another.

Evaporation, with its close relationship to weather, is a very important part of the hydrologic cycle. More than 80 per cent of all evaporation occurs from the ocean surface, most of it from the warm waters in the tropical and equatorial zones. Large areas of the ocean have an annual evaporation exceeding 200 cm per cm² (see Figure 4.22). They are found in the trade-wind belts, 5 to 10 degrees away from the equator, and in the warm ocean currents off Malagasy and Virginia. Evaporation from the continents comes from the open water of lakes and streams, from snow, from bare soil, and from growing plants.

Precipitation in all its forms returns the water to the earth's surface. Here it may flow through stream and river channels to the ocean; it may percolate into the soil to become part of the groundwater; it may

THE HYDROLOGIC CYCLE

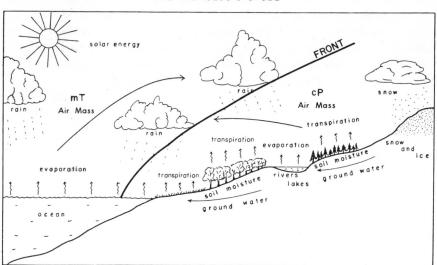

FIGURE 12.2The hydologic cycle. (Published by permission of John R. Mather. From Mather, John R.: The average annual water balance of the world. *Proc. Third Water Resources Symp. of the Amer. Water Resources Assn.,* Banff, Alberta, June 1969).

penetrate even deeper into the underlying rock formation; or it may return to the air by evaporation or evapotranspiration. Eventually all the rain, except that which reevaporates, finds its way back to the ocean, but the rate of movement varies. Some may become locked for years in mountain glaciers or in large lakes or underground rocks or in the tissues of plants.

Because the moisture below ground level tends, in the long run, to reach a balance, precipitation tends either to pass back into the atmosphere by evaporation and evapotranspiration, or to flow through streams and rivers to the oceans. The proportion that is returned to the ocean varies with the region. On glaciers and ice caps, such as Antarctica, evaporation is almost zero, and runoff nearly equals precipitation. Desert areas are examples of the other extreme, and other near-arid regions are similar. Water flowing out of the area is balanced against the inflow from surface and underground streams, and evaporation and precipitation are approximately equal. Considering larger areas, the evaporation from most of the continents amounts to 60 to 75 per cent of the precipitation, and over the dry continent of Australia the evaporation is nearly 90 per cent of the precipitation.

The world distribution of precipitation and of evaporation (see Figure 4.22) permit an examination of the water balance of the globe. One such study gave the mean precipitation and evaporation for 10° latitude belts from 80°N to 80°S. The results are found in Table 12.1. Part of the variation among belts arises because of the differences in area and in the land/sea ratio. The results, particularly those over the ocean, also show the climatic zones: the equatorial trough, the subtropical anticyclones with their low rainfall and high evaporation, the zone of westerlies with their traveling storms, and the cold polar regions.

By means of the data in Table 12.1, we can calculate the flow of moisture across the different latitude circles. Thus the flow of moisture across 40°N latitude is 1.81×10^{19} cm^3. This represents a flow of latent heat of 1.1×10^{22} cal or an average flow of 3.0×10^{19} cal day^{-1}. This value is higher by 25 per cent than the value found in Table 1.2 for 42.5°N. The difference, although large, is not unexpected in view of the many assumptions and approximations made in both studies. The fact that the results obtained by two widely different methods are in so close agreement leads one to have confidence in the methods and in the results found in Tables 1.2 and 12.1.

The data in Table 12.1 bring out clearly the sources and sinks of our atmospheric moisture. In every band from 40°S to 40°N the evaporation exceeds the precipitation, with the exception of the band from equator to 10°N. In this band, the precipitation is $\frac{1}{3}$ more than the

TABLE 12.1 Estimates of the Annual Precipitation (P) and Evapotranspiration (E) from Land and Water by Latitude Zones (after Mather[a]) (Units 10^2 km³ or 10^{17} cm³)

Zone	P Ocean	E Ocean	P Land	E Land
70 to 80°N	19	22	6	4
60 to 70°N	32	27	48	22
50 to 60°N	112	60	76	42
40 to 50°N	164	126	85	58
30 to 40°N	187	281	85	61
20 to 30°N	190	385	89	53
10 to 20°N	370	492	94	71
0 to 10°N	714	479	160	108
10 to 0°S	440	502	202	115
20 to 10°S	309	568	114	78
30 to 20°S	231	469	52	40
40 to 30°S	301	380	24	17
50 to 40°S	372	226	6	4
60 to 50°S	275	118	2	1
70 to 60°S	88	47	3	2
80 to 70°S	9	7	15	9
Totals	3813	4189	1061	685

[a] From Mather, John R. The average annual water balance of the world. *Proc. Third Water Resources Symp. of the Amer. Water Resources Assn.*, Banff, Alberta, June, 1969.

evaporation, a result of the trade winds which carry moist air from both directions into the area. A similar excess of precipitation is found near the subpolar lows, with once again the balance being maintained because of the flow of moisture from the subtropical highs.

Although the moisture in the soil may be considered constant when dealing with a period of a year, changes that are significant for plant growth occur during the course of the year. After a wet period, the soil moisture reaches *field capacity* and any additional moisture flows away. During a dry period, the soil moisture is removed by evapotranspiration. This will continue, unless further precipitation replenishes the loss, until it falls to the point where the vegetation cannot obtain sufficient moisture and, hence, reaches the *wilting point*. This is the beginning of a drought period.

When the returning rains are gentle, the water penetrates the soil and the moisture content starts to rise again toward field capacity. During heavy rains, the movement into the soil is too slow to absorb that which falls and thus some is lost to the area through surface flow. Normally soil moisture returns to field capacity during the cool season, at least in the temperate zone.

The hydrologist is concerned with providing adequate water supplies for power dams, for irrigation, for urban use, and for similar purposes. He also tries to combat floods, erosion, and water pollution. The domains of the meteorologist and the hydrologist meet where the rain hits the earth and again when the water evaporates from the earth and water reservoirs. The basic data for the hydrometeorologist are the statistics of precipitation. He needs to know the total rainfall over the area even more accurately than the meteorologist. When the rain gages are too far apart to provide adequate information for the total area, as they must be in mountainous regions, then the hydrologist must extrapolate his data as best he can into these unknown areas.

One of the basic problems of hydrometeorology is the estimation of the flood potential of a specific stream. From the answer comes the decisions about the heights of levees, bridges, etc. The hydrologist collects data on the heights of floods at one or more observing points on the stream. These he relates to the amount of rain that has fallen and the amount of snow that has melted in the drainage basin. In computing the flood potential, he deducts from them the amount of water that is stored in lakes and reservoirs and that percolates into the ground. His data on stream flow show that the volume of water rises rapidly to a maximum and then tapers off slowly. The lag between the time of the rain and the time of maximum flow increases, and the increase of height above the normal flow decreases, with distance downstream.

From a series of observations, the hydrologist derives a *unit hydrograph* to give the flow from a unit fall of rain on the river basin in a unit time (that is, one inch in one day). With this as a tool, he is able to estimate the flow pattern for any rain on the area. From the records of the area, he can make an estimate of the probable maximum rainfall, and from this the probable maximum flood for the observation point.

Winter precipitation in areas where the ground freezes presents a different problem to the hydrologist. When the ground is frozen, little percolates into the ground, although some may become locked in ice. The resulting runoff in autumn differs only slightly from the total rainfall. The spring runoff is complicated both by the snow melt and by the effect of ice in the rivers, which interferes with a smooth flow. With information on the probable temperature and the state of the snow cover, an estimate can be made of the addition the snow makes to the flow. The effect of the river ice is more difficult to assess, particularly if the river flows toward higher latitudes. The northern section may remain frozen while warm air lying over the upper drainage basin melts the winter snow cover rapidly.

When the river rises in the mountains, the spring flood is fed by the melting of the snow fields at high altitudes. With a gradual rise in temperature, the spring runoff is spread over days or weeks. When, as sometimes happens, a cool period with little melting is followed by sudden warm rains, rapid melting over wide areas results in widespread flooding.

The hydrologist includes in his calculations for river levels in spring the runoff from the snow melt as well as from the rain. The former cannot be evaluated accurately. The energy to melt the snow may come from the short-wave radiation from the sun, from the long-wave radiation and conduction of heat from overlying warm air, from the warm rain that falls into it, or from the latent heat, as water vapor from saturated air condenses on the snow. By experience plus a study of the weather records, the hydrometeorologist is able to estimate the probable runoff from the snow pack and, therefore, the intensity of the flood on the river.

Evaporation is a common problem for the hydrologist and the meteorologist. The meteorologist sees it as a source of water vapor and also as a means of storing energy. The hydrologist seeks methods of computing the vapor loss from the growing crops and from his reservoirs and irrigation ditches. These losses are the ones that must be estimated in calculating the amount of water that must be in storage in the reservoirs at the end of the wet season to meet the demands during the dry season. Studies of evaporation have been made by hydrologists, such as Thornthwaite and Penman, and equations have been obtained by which to calculate the evaporation. The basic data for these equations have been gathered by meteorologists and climatologists.

Thus the work of the hydrologist is closely linked with the weather. He could often make effective use of reliable long-range forecasts. This would be particularly true in early spring, and again when he must estimate what demands there will be on water during the heat of the summer. He is, and will continue to be, in close contact with the meteorologist, seeking whatever help is available.

12.5 *Weather and Health*

During the course of his history, man has proved able to adapt himself to life in most of the earth's climates and to combat the unfavorable features of these climates. Prehistoric tribes learned to live in the hot, humid areas of Africa and Brazil, in the hot deserts of Arabia and Australia, in the damp raw weather of Tierra del Fuego, and in the dry cold areas of northern Canada and Siberia. The tribes within the Arctic Circle made and adapted clothing to combat the extreme cold, but the people on the southern tip of South America were able to exist without much covering.

Adapted to the climate, men are still affected by the weather, that is, the day-to-day changes, in numerous ways. For example, the incidence of colds and other infections of the respiratory tract rises rapidly in periods of rapid weather changes. The body does not adjust immediately to the changes that occur and, hence, is in a condition susceptible to infection. Some medical geographers have concluded that the stresses induced by the changing weather have been one of the causes for the advancement of civilization during the past 8000 years. These stresses, they say, have forced man to struggle to combat them and also have provided periods when consideration could be given to conquering them. The inhabitants of tropical areas lacked the periods of stress; the people around the pole lacked periods of relief. Civilization developed in the temperate zone where both are present. Such an hypothesis is difficult to prove, but it is possible to advance good arguments to support it.

Measuring the comfort of the individual under varying weather conditions has been studied extensively. The temperature gives a first approximation, but a first approximation only. The wind is another weather element that contributes to comfort or discomfort. Strong winds with low temperatures make a man feel cold because they rapidly replace the warm layer of air about the body with fresh cold air. *Wind chill* is the rate at which a body at a temperature of 98°F cools under different temperature and wind conditions. The equations to compute wind chill are complicated, and the results are not directly applicable to a man who clothes himself in a variety of garments. A rough-and-ready evaluation of the effect of wind can be made by assuming that an additional mile per hour of wind is equivalent to an additional degree Fahrenheit drop in temperature. Thus, a situation with a wind speed of 30 mi hr^{-1} and a temperature of 0°F feels equally cold as one with a wind speed of 5 mi hr^{-1} and a temperature of −25°F.

The discomfort at high temperatures depends on the relative humidity as well as the temperature. The body loses surplus heat by the evaporation of perspiration, the rate of which depends on the air temperature, the wind speed, and the relative humidity. With high temperature and high relative humidity, the body is unable to lose surplus heat rapidly enough, and a heat stroke may result. Again, it is possible to arrive at a figure that gives a measure of this discomfort.

Both measures discussed attempt to measure human reactions by means of meteorological variables, but both omit the most important variable, the human animal. A man, transported suddenly into another environment, finds the weather harder to bear than those who reside there. If he is healthy, even a week in the new surroundings will help him to become adjusted or to "get used to it." Any attempt to measure the comfort or discomfort of the weather must be based on the "average" individual and cannot be put into precise terms.

Assistance in counteracting the effects of the weather can be provided by proper clothing and food. Because, in warm weather, comfort is supplied by evaporation, methods to increase it are desirable. Light, porous clothing allows the wind to penetrate to the body, removing the moist, body-warmed layer of air and permitting more evaporation. The Arabians, traveling in the desert, wear loose folds of clothing to permit ventilation and increase comfort. The light color also adds to the comfort because the clothing absorbs less of the intense solar radiation. Foods that supply moisture (for instance, salads and soups) are better during hot spells of weather, and heat-producing starchy foods should be eaten in small quantities.

For cold windy weather, clothing that is wind-resistant and that decreases the ventilation of the body is the most suitable. Absolutely airtight clothing is not desirable because it keeps the body moisture from escaping, but the clothing developed by the Eskimos is very nearly airtight. On the other hand, the normal overcoat provides poor protection in windy weather since it permits a flow inside its covering. Because condensed moisture, which is a good conductor of heat, lying next to the body allows heat to be dissipated rapidly, dry clothing is highly desirable. Norwegian fishermen used the low conductivity of air to keep the body warm by wearing a string open-net vest next to the body when fishing in cold waters. Suitable foods for those who must be out of doors in the cold should supply heat and energy to combat the rapid heat loss. The need for water is considerably reduced but not negligible.

12.6 Weather and Air Pollution

One problem of health that is accentuated by the weather is the effect of air pollution. Impurities, such as smoke, dust, plant pollen, etc., have been passing into the atmosphere since before the dawn of history. Turbulence in the air has reduced the concentrations to small values; gravity has caused the heaviest particles to settle to the ground; and rain and snow have cleaned the air by carrying some impurities to the earth. The air has thus remained relatively pure until recently.

With the growth of civilization in the temperate zone, coal-burning fires resulted in increased pollution, reaching at times undesirable concentrations. Three hundred years ago, John Evelyn, a Fellow of the Royal Society of London, reported on the smoke concentrations over London. The bad effects have increased rapidly during the past century with the increase of manufacturing. The polluting material consists not only of unburned carbon particles but also of a number of noxious gases, that is, carbon monoxide, sulfur dioxide, sulfur trioxide and its derivative sulfuric acid, oxides of nitrogen, aldehydes, fluorides, and radioactive gases. Manufacturing plants, fires for heating, and automobile exhausts are all guilty of adding to the pollution.

The effects of these substances are much less when the concentration is kept low by turbulence. With a stable situation, turbulence is low, and the concentrations continue to increase as the sources of impurities continue to supply more of their products to the atmosphere. Normally, an inversion develops about sundown and remains until the early morning (Section 6.9 and elsewhere). If the air is still and the night long, the pollution can be high by morning. The turbulence during the day will mix the lowest layers of air and, thus, reduce the concentrations.

In a location where factors other than nighttime radiational cooling tend to produce inversions, they may persist for several days, permitting the concentrations to increase to unhealthy values. A valley bottom provides a situation favorable to the development of severe pollution. The winds tend to be light, the cold air moves down the valley sides, and the sun's heat is less effective than on the sides of the valley. Pollution of this type occurs in a number of mountain valleys, such as the Columbia River Valley of British Columbia, where mining and smelting are being done.

Even more significant are those meteorological situations where inversions persist during the day. Normally inversions are found in the lowest layers of a polar continental air mass at or near its source. When an anticyclone moves over a city, the pollution can increase rapidly. The solar heating during the day may decrease the stability somewhat, but usually not sufficiently to mix the polluted air with the clean air above the inversion. Therefore, the pollution continues until the weather situation changes. Generally, this describes the situation that has made the London fogs so bad. In winter, a high-pressure area of polar maritime air sometimes settles over England. In the inversion, the noxious gases of London used to reach levels that caused serious discomfort or even death. Pollution control measures, prohibiting the emission of noxious gases into the air have reduced the danger in recent years.

Both adverse factors combine when a cold anticyclone settles over a hilly region in which valleys are heavily industrialized. At Donora, in a mountain valley in southwest Pennsylvania, from 26 to 31 October 1948, the inversion was limited to the lowest 1000 ft by a stagnant high-pressure dome. The poisonous gases were not carried away, and pollution levels increased during the six-day period; many people became ill and 20 died.

The cool waters off the California coast provide another situation that leads to serious pollution. Tropical maritime air on the west coasts of continents is normally stable, with a cold ocean current increasing the stability. The cars and industrial plants in the Los Angeles area supply poisonous gases, while the light winds and the barrier of the mountains to the east restrict their dilution. Because Los Angeles is in an area of few frontal passages, there is seldom a change of air mass and, consequently, the pollution dissipates slowly.

The meteorological phenomena that contribute to pollution cannot be controlled and must be accepted. To reduce pollution, emission of impurities must be controlled or prevented in those regions and at those times when they can accumulate to dangerous proportions. Many cities have begun to recognize the hazard and enforce strict laws to reduce the quantity of impurities that pass into the atmosphere.

12.7 Weather and Other Human Activities

Weather changes affect many other businesses and occupations. Construction is hindered by wet or stormy weather. Construction engineers have developed a number of methods by which they can continue building, although at a slower pace, during unfavorable weather. Some logging operations are delayed until frost has made the swamps safe for travel; a light snowfall can be beneficial but excessive snow can slow operations to a standstill. Most sporting events are subject to the whims of the weather. Yachting demands a moderate wind, and baseball and tennis can be played only when there is no rain. Football can be played in almost all kinds of weather, but suffers financially in inclement weather, and even a football game must be called off in dense fog.

Surface transportation is sensitive to certain types of weather. Poor visibility caused by fog, snow, or dust reduces train movement to a certain extent and road traffic even more. The worst hazard for road travel is the occurrence of freezing rain which makes the highways icy. At times, the rain may not be freezing as it falls, but will freeze on impact with the frozen ground. Packed snow will also make a dangerous surface at times. This effect, along with the hiding of the edge of the roadway and the guidelines from the driver, causes traveling in a snowstorm to be at a very much slower pace than normal. Wind not only causes blowing snow, but sometimes high gusty winds may carry a car out of its proper lane into oncoming traffic or off the road and into the ditch. This can be particularly serious with a large trailer attached to the rear of an automobile. Snow and extreme cold both interfere with train schedules, although the disruption is not so extreme as with road traffic.

A major problem in power transmission is the occurrence of an ice storm. The coating of ice can seriously overload the power lines to cause some to break and supporting pylons to collapse. Some protection can be given by putting additional power through the wires to keep them warmer and hinder the formation of ice. The surge of power initiated by a lightning bolt will burn out transformers unless protected from lightning, and interrupt the service. The demand for power, too, is affected by the weather: additional power is required during cold windy weather, and modern air conditioning has increased the need for power during humid summer days. Even a sudden clouding over of the sky prompts housewives to turn on lights in sufficient numbers that the demand for power increases with the darkening of the sky.

The tourist industry is very sensitive to weather variations, particularly on weekends and holidays. Much of the beauty of the landscape, such as the trees and the flowers, requires the cloudy days with their life-giving rain. But the tourist wants to see the beauties and photograph them against a background of a blue sky flecked with some fleecy white cumulus clouds. A sequence of wet weekends can be a financial disaster to people who depend on tourists for their income.

The foregoing are some of the ways by which weather affects the activities of mankind. The trend to industrialization accompanied by the move of a large proportion of the population into cities has reduced greatly the dependence that the human race has had in the past on the weather. We listen to the weather forecast regularly in the morning, but on very few days does it concern the average individual vitally. Nevertheless a frost in Florida can increase the cost of oranges across the nation; a wet harvest season can reduce the quality of the crop and increase the price of good food; the drought of the 1930's was a contributing factor in delaying the recovery from the depression; and, to go farther afield, a crop failure in India can result in increased sales of North American wheat. It may be that the current weather forecast has little significance to many, but the weather in our district and in other parts of the world touches our lives in many unknown ways.

CHAPTER 13

CLIMATIC CHANGE

13.1 *Temperatures in Past Eras*

The balance between the energy coming from the sun and that passing from the earth to outer space has been discussed in several sections of this text, particularly in Chapter 1 and Sections 5.4 and 5.5. The treatment has been based on the assumption that the two flows balance when considered on an annual basis or longer, resulting in temperatures that vary little from year to year. A careful examination of records kept during the past 200 years shows that the conclusion is not far wrong.

It is simple to consider that our climate has been uniform through the years, but scientific investigation has shown that over long periods of time the climate has changed. Geologists have shown that the earth has passed through cycles of the order of millions of years. During most of the time in each cycle the temperatures were much above our present temperatures, and tropical vegetation has grown under the shadow of the poles. These warm periods have been interrupted by four widely separated epochs when glaciers advanced out of the polar regions and into the so-called temperate latitudes. The last of these glaciers had four separate maxima; the last of these waves retreated from northern Europe and northern North America about 10,000 years ago. Climatologists consider that we are still not truly out of the ice age, but are merely in an interglacial period similar to other periods of temporary glacial retreat.

Other scientists have studied the changes of temperature during the 10,000 years since the last glacial recession, and have concluded that there have been cool periods and warm periods, wet periods and dry periods in the intervening years. The era about 5000 B.C. was very warm, a climatic optimum, after which the earth cooled again. The 10th and traveled across the North Atlantic and established a colony in Greenland, 11th centuries, when Eric the Red, Leif Ericsson, and their companions were relatively warm. Thereafter, the weather in western Europe turned stormy and cool, reaching a minimum of temperature in the 14th and 15th centuries.

The cycles of warm and cold thus far considered have been deduced by indirect evidence because there were no carefully kept records. Some individuals began to keep weather records during the 18th century, and many more began in the nineteenth. Figure 13.1 shows changes of temperature discovered from these records for a few of the regions of the earth. These graphs are obtained by a method common in this type of study, the method of moving averages. Each point represents the average annual temperature for a period of five consecutive years.

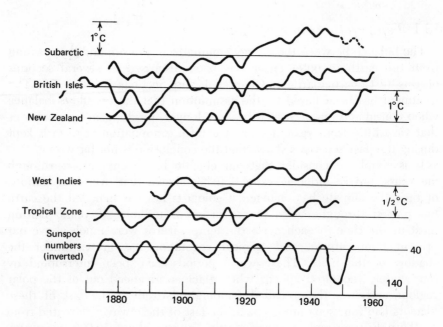

FIGURE 13.1 Temperature fluctuations. Running five-year annual mean departures from the mean for 1901 to 1930, with the inverted sunspot numbers. (Published by permission of G. S. Callendar. From Callendar, G. S., 1961: Temperature fluctuations and trends over the earth. *Quart. J. Roy. Meteor Soc.*, 87, 4, 5.)

The temperature changes of the past, both the recent and the geological past, give rise to a question on the meaning of a "normal" temperature. Considering the time period of the geologists, the "average" temperature is much higher than at present. Even the changes during the present interglacial period make it difficult to say what is "normal." In practice, climatologists use the average for a recent period of approximately 30 years length as the normal, but they recognize that this normal does change, even during a lifetime.

13.2 Some Effects of Temperature Changes

The variations shown in Figure 13.1 appear small, with the swing from the coldest five years to the warmest five years being under 2 deg C. Nevertheless, the change is sufficient to affect farmers who discover that they are able to, or forced to, alter their crops as a result of higher or lower temperatures. Research scientists studying glaciers in Iceland, Europe, western North America, and elsewhere have been able to discover that the swings in temperature illustrated in Figure 13.1 have been sufficient to cause glaciers to retreat and to advance. of Canada. The migrations of certain kinds of fish have changed with are now growing farther north than previously in the Ungava Peninsula As a result of the warming trend during the past 40 years, birch trees the temperature cycles. Thus, even minor changes like the ones shown in Figure 13.1 can alter plant and animal life on our planet.

The changes in climate during historical times have had their influences on certain events in human history. For example, the settlement in Greenland flourished during the warmth of the 11th century, and then disappeared when colder weather returned three centuries later. Some of the migrations of the Teutonic tribes at the time of the decline of the Roman Empire were, it is believed, forced by the lack of rain in their normal pasture lands. Some historians attribute the vigor that produced the Greek and Roman civilizations to the cooler, more stormy weather of the pre-Christian era in comparison with that of later centuries. Other examples of the influence of climatic change can also be found.

13.3 Spells of Weather

We discussed in Section 9.3 the change in weather that occurred The circulation in the upper atmosphere stagnates for several weeks such short-period variations with the general variability of the weather. in North America between November and December 1963. We associate somewhat out of phase with what we consider normal, and so we have a "spell of weather" that is unusual for that time of the year. It is expected that these spells average out over a period of a year or two. The changes in the temperature noted in Figure 13.1 are indicative of changes in the general circulation for longer periods for which we would like to have an explanation. It is probable that some of the causes for the longer-period changes may operate in the deviations for a few weeks, but it will require more study to determine to what extent this is true.

13.4 *The Heat Balance*

Figure 13.1 shows that the mean temperature has changed during the period of meteorological observations, and the data from preceding centuries show greater variations. These contradict the hypothesis that there is a balance among the terms of the heat balance equation. Going from effect to cause, the swings in temperature give evidence that the terms may not balance for centuries at a time. It is impossible to attempt to calculate the correction term because this would be a relatively small difference between two large terms. Nevertheless it is useful to examine the changes that may have occurred in the past or that may be occurring at present that would throw the equation out of balance. With any change in one term, adjustments will occur in the remaining terms with the result that the equation may balance once again, probably with a shift in the climate of the earth.

Some people have postulated that the changes have resulted from extraterrestrial phenomena or from changes in the earth. For instance, they have wondered whether the eccentricity of the earth's orbit around the sun has changed, or if the earth's axis has shifted relative to the continents. The proof or refutation of each of these hypotheses is a problem not for the meteorologist but for the geophysicist. Even if they can be used to explain the geological eras of the remote past, they cannot be considered as valid causes for the retreat of the glaciers 10,000 years ago and for the variations in climate for which we have evidence since that retreat. As far as man has been able to determine, the relative positions of earth and sun and of poles, land masses, and oceans have been constant or near constant during the past 10,000 years.

13.5 *Causes for a Changing Climate—The Solar Constant*

The major term in the heat balance equation is the heat that the earth receives from the sun. Many scientists have pondered on the effect that a variation in the solar "constant" might have on the earth's climate.

Heretofore the solar constant has been estimated only with approximations that make comparisons difficult. A period of sunspot activity is considered to be a time of relatively high solar energy output. The cycle of sunspot activity has long been known, and a careful record goes back many centuries. Figure 13.1 gives the inverted sunspot curve for comparison with temperature changes. It appears that this inverted sunspot curve and the curve of temperature in the tropics moved together until 1920, after which the relationship decreased.

A warming sun will affect two factors that have been discussed in this book, the equator-pole temperature gradient and the land-sea contrast. Regarding the first, an increase of the solar constant will be noticed at the equator first. This will result in increased evaporation, and also in a greater temperature contrast between equator and poles. Poleward advection of heat will increase, a change that will be made only through an increasing intensity of the moving cyclones and anticyclones. The increase in evaporation will result in more clouds and greater precipitation over the temperate and polar zones. The clouds, by reflecting solar heat, will keep polar temperatures low. If they remain sufficiently low, the precipitation in the form of snow will not all melt in the summer season. Snow and ice will accumulate and will start moving southward to initiate another ice age. Thus did G. C. Simpson, an English meteorologist, reason in the 1920's to deduce that a hot sun might be the cause of an ice age.

The second effect of variable sum is a change in the land-sea contrast. To understand something of its complexity, assume that the departure of the incoming solar energy from its normal value follows a sinusoidal curve, that is,

$$e = E \sin \frac{2\pi}{t_0} t$$

where e is the departure from normal at time t, E is the amplitude of the wave, and t_0 is the period of the wave. This is illustrated in Figure 13.2.

The change in solar energy will result in a change in the surface temperature, both over the land and over the sea. They, too, will follow a sine curve approximately, but with a lag in comparison with the solar energy. The curves will be similar to the annual temperature curves, with the lag being different over the two surfaces just as the annual curves show different lags (see Section 5.6 and Figure 5.5). If T_L gives the departure from the normal temperature of the land resulting from

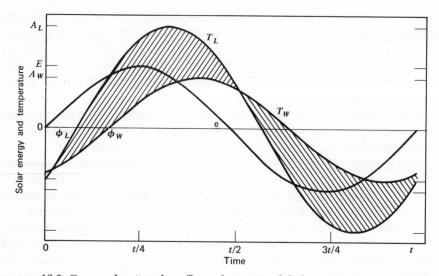

FIGURE 13.2 Curves showing the effect of a sinusoidal departure of solar energy from normal (e) on the temperature of the land (T_L) and ocean (T_W). The amplitude of the terrestrial temperature curve is assumed double, and the lag one-half, of the corresponding values for the oceanic curve.

the changing value of e, and T_W the corresponding value for the ocean, then approximately

$$T_L = A_L \sin\left(\frac{2\pi}{t_0} t - \phi_L\right) \tag{2}$$

and

$$T_W = A_W \sin\left(\frac{2\pi}{t_0} t - \phi_W\right) \tag{3}$$

where A_L and A_W are the amplitudes and ϕ_L and ϕ_W the phase angles, respectively, of the curves over land and over water (Figure 13.2). From the foregoing discussion, and from the discussion of Section 5.6, we see that

$$A_L > A_W \quad \text{and} \quad \phi_W > \phi_L$$

As stated in Section 7.12, pressures tend to be high over cold areas and low over warm areas. Therefore, the temperature difference between land and ocean controls somewhat the pressure distribution. The effect of the change in the solar constant is a change in the water-land temperature difference given by

$$T_W - T_L = A_W \sin\left(\frac{2\pi}{t_0} t - \phi_W\right) - A_L \sin\left(\frac{2\pi}{t_0} t - \phi_L\right)$$

In drawing Figure 13.2, it was assumed that $\phi_W = \pi/6$ and $\phi_L = \pi/3$ which, based on the annual curve, seem reasonable values. The ratio of the amplitudes, A_W/A_L was assumed equal to $\frac{1}{2}$. Under these assumptions, the temperature difference (shown by the shaded areas) follows closely the curve of solar energy. It would appear, then, that the land-to-ocean pressure difference would be greatest at the time of maximum solar heat and would be minimum at the time the sun is coldest. Studies of pressures have shown that oceanic pressures tend to be high at times of maximum sunspot activity and low at times of sunspot minimum.

The many assumptions in the foregoing analysis must be confirmed; the relationship between observations and analysis may be fortuitous. When pressures are high over the north Pacific Ocean, northwest winds will develop over the western half of the continent, a trough will form along the Mississippi Valley, and southwest winds blow along the eastern seaboard. Immediately we can see that the change will produce different effects in different parts of the continent. In the west it will be cool and dry; in the east, warm and wet.

Evidence has been given that some of the suggested results do occur. Tannehill (1947) and others have shown that when the pressure at Portland, Oregon (and presumably the Pacific) is high, the average precipitation over continental United States is low, and vice versa. Something of this effect is indicated in Figures 13.3 and 13.4. They, from work by Kincer,[1] show that pressures were low in the state of Washington and high over the interior of the continent during the wet decade of 1906 to 1915. During the dry decade of 1930 to 1939, the reverse held true, with high pressure over much of the mountain area of the United States and low pressure over the Mississippi Valley.

13.6 *Causes for a Changing Climate—The Albedo*

One influence that can cause a climatic change is the albedo. An increase in this quantity would produce effects similar to the ones from a decrease in the solar constant, because in both situations the absorbed heat decreases.

The albedo of the earth is not constant. A sudden increase in the albedo for the Northern Hemisphere occurred in 1883 after a volcano erupted and blew up the island of Krakatoa in the Pacific. The explosion was so great that dust was carried high into the stratosphere, and some remained for several years in the stable layer at the top of the mesosphere. Sunsets for the next few years were more colorful because of this dust,

[1] Kincer, J. B., 1946: *Our changing climate.* Trans. American Geophysical Union.

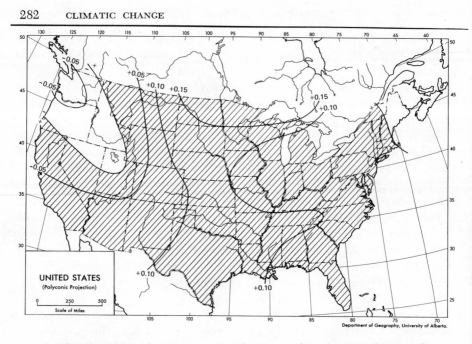

FIGURE 13.3 Departure of station pressure from normal in the United States during the wet decade, 1906 to 1915 (after Kincer). (From Kincer, J. B., 1946: Our changing climate. *Trans. Amer. Geogr. Union.*

which also reduced the solar radiation reaching the earth's surface. Mean temperatures at many places were reduced (see Figure 13.1) until after the turn of the century. Some scientists have suggested that the glacial periods of the past followed periods of mountain building and volcanic activity and that the cold periods were a consequence of the volcanic action and the resulting additional dust in the atmosphere.

There is reason to wonder whether mankind is inadvertently modifying the climate and weather by changing the albedo. The increase in the atmospheric haze that comes from pollution will result in a greater loss of heat by reflection. The condensation trails from our stratospheric aircraft increase the amount of cloud in the sky and, hence, the albedo of the earth. The change may be small, but even a small change can affect the radiation balance for the globe, and an adjustment elsewhere must follow the changing albedo.

13.7 *Causes of a Changing Climate—Long-wave Radiation*

If the amount of absorbed solar heat remains unchanged, the flow of heat outward from the earth must also remain unchanged. But changes in the atmosphere may alter the flow and with it the temperature.

As described in Chapter 5, the outward flow of radiant heat from the surface of the earth through the atmosphere to space is complex. Some passes directly through the atmospheric "window" near 9μ. Some is absorbed and reradiated. Changes in the moisture of the atmosphere, either vapor or droplets in clouds, result in a change in the flow pattern. These changes tend to repeat themselves in the annual cycle, and so they cannot explain a changing climate unless some other influence alters the mean moisture content of the air. As mentioned in Section 13.5, this could occur with an increased solar constant.

The other gas of significance is carbon dioxide (Section 1.3). It has a cycle similar to the hydrologic cycle (Section 12.4) in which the vegetation and ocean play parts. Presumably, the proportion of carbon dioxide in the atmosphere has remained nearly constant through the ages, but this may no longer be true. The development of manufacturing and other activities of mankind have resulted in the burning of our fossil fuels—coal and oil. Carbon dioxide from the burning has passed into the atmosphere, increasing slightly its relative amount.

The increased carbon dioxide will absorb and will reradiate more

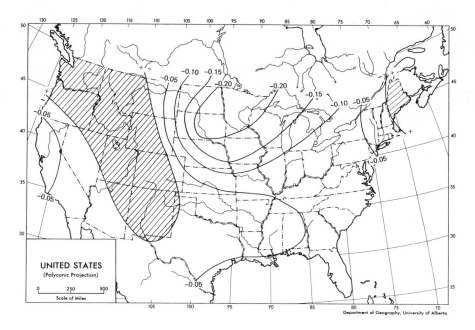

FIGURE 13.4 Departure of station pressure from normal in the United States during the dry decade, 1930 to 1939 (after Kincer).

of the long-wave radiation from the earth, upsetting the normal flow. Reradiation back to the earth will increase the surface temperature. Thus a new balance will be established with higher temperatures. The dusts and smoke also influence the flow of the long-wave radiation.

In Sections 13.5 to 13.7 an attempt has been made to interpret the effects when one specific meteorological element changed. But the discussion has shown that as one element changes another element also changes; and the latter also affects the climate of the earth, sometimes augmenting the effect of the first change and sometimes counteracting it. Also a change in one element is not equally significant in every part of the earth, and it may cause the climate to change in opposite directions in two different places. Thus a ridge of high pressure over the Pacific brings colder drier weather to the western plains and warmer wetter weather to the Atlantic coast. These interrelationships among the many variables provide complexities that make it difficult or even impossible to determine the total effect.

13.8 *Man-made Changes in Climate*

The discussion found in Sections 13.6 and 13.7 have suggested that human activity has inadvertently caused some changes in the weather and climate of the earth. Mankind has not stopped here, but has considered the possibility of altering the climate in a planned manner. Already this is being done on a small scale. It began in prehistoric times when man built a fire in a cave and so changed the temperature pattern there. But most of such activity has come during the past fifty years. With frost protection devices, and with cloud seeding to make rain or to prevent hail or to reduce the damage from a hurricane, mankind is attempting to change the climate in a small area.

Larger scale projects have been suggested. One of them is to cover the Greenland ice cap with dust which, absorbing the solar radiation, would cause the ice to melt and disappear. Another suggested project is to increase the flow of warm water into the Arctic Ocean and thus melt the sea ice. Both of them would alter the climate of much of the Northern Hemisphere, but scientists do not agree on the ultimate result.

Some scientists claim that an open Arctic Ocean would absorb more solar heat than the ice pack. This would raise the temperature of the surrounding land, and reduce the northward temperature gradient and, therefore, the vigor of the storms in the middle latitudes. The flow of water vapor from the equatorial regions would be reduced, and the desert areas of the subtropics would expand. In return for a milder climate of the polar regions, many other parts of the hemisphere would suffer.

Other scientists have proposed that an open Arctic would initiate an ice age. The open Arctic would, they claim, provide considerable moisture to the artic air masses. This would be deposited as snow on the Canadian north and on northern Siberia. The amount would be so great that the summer heat would fail to melt it, and it would accumulate. Finally the depth of snow would result in glaciers moving southward as in the past glacial ages.

It is impossible to say with confidence what effect an open Arctic would have on world climate. Yet the situation is not hopeless. Electronic computers are designed to help solve complex problems, and already they are proving valuable in solving problems of the current weather. As meteorologists become competent in dealing with the actual weather, they anticipate that they can insert the problem into the computer and solve the hypothetical situation that would result with the changes that are proposed. The answer from the computer may not be the correct answer, but it should provide man with an answer in which he will have more confidence than he has with any answer advanced thus far.

CONVERSION FACTORS AND CONSTANTS

1 cm	= 0.3937 in.
1 km	= 0.621 stat mi
1 micron	= 10^{-4} cm
1° lat	= 111.1 km
	= 69.06 stat mi
	= 60.0 naut mi
1 cm²	= 0.155 in.²
1 km²	= 0.386 stat mi²

1 cm³ = 0.0610 in.³

1 m sec⁻¹ = 2.237 mi hr⁻¹
1° lat day⁻¹ = 1.286 m sec⁻¹

1 kg = 2.205 lbs
1 gm cm⁻³ = 62.4 lb ft⁻³

1 in.	= 2.54 cm
1 stat mi	= 1.609 km

1 in.²	= 6.45 cm²
1 ft²	= 929 cm²
1 stat mi²	= 2.59 km²
1 in.³	= 16.39 cm³
1 ft³	= 28320 cm³
1 mi hr⁻¹	= 0.447 m sec⁻¹
1 knot	= 1 naut mi hr⁻¹
	= 0.515 m sec⁻¹
1 ft sec⁻¹	= 0.681 mi hr⁻¹
	= 30.5 cm sec⁻¹

1 lb = 454 gm
1 lb ft⁻³ = 1.602×10^{-2} gm cm⁻³

1 mb $= 10^3$ dynes cm^{-2} 1 mm mercury $= 1.333$ mb
    $= 0.750$ mm mercury   1 in. mercury $= 33.9$ mb
    $= 0.0295$ in. mercury
    $= 0.0145$ lb in.$^{-2}$

1 standard atmosphere
        $= 1013.25$ mb
        $= 760$ mm mercury
        $= 29.92$ in. mercury

1 erg $= 1$ dyne cm
        $= 2.389 \times 10^{-8}$ cal

1 cal $= 4.187 \times 10^7$ ergs

1 langley $= 1$ cal cm^{-2}

1 radian $= \dfrac{180°}{\pi} = 57.3°$

For dry air

c_p (specific heat at constant pressure) $= 0.240$ cal gm^{-1} deg^{-1}
c_v (specific heat at constant volume) $= 0.171$ cal gm^{-1} deg^{-1}
R (gas constant) $= 2.87 \times 10^6$ ergs gm^{-1} deg^{-1}

For water, specific heats.

For ice	$c = 0.50$ cal gm^{-1} deg^{-1} at 0°C
For liquid water	$c = 1.00$ cal gm^{-1} deg^{-1}
For water vapor	$c_p = 0.44$ cal gm^{-1} deg^{-1}
	$c_v = 0.33$ cal gm^{-1} deg^{-1}
Latent heat of vaporization	$= (597.3 - 0.566$ T°C$)$ cal gm^{-1}
Latent heat of fusion	$= 79.7$ cal gm^{-1} at 0°C
R (gas constant)	$= 4.615 \times 10^6$ erg gm^{-1} deg^{-1}

Polar radius of earth $= 6357$ km
Equatorial radius $= 6378$ km
Angular velocity of the earth, Ω, $= 7.29 \times 10^{-5}$ radians sec^{-1}
Acceleration due to gravity, g, at 45° latitude $= 980.616$ cm sec^{-2}
Stefan-Boltzmann constant, σ, $= 8.13 \times 10^{-11}$ cal cm^{-2} °K^{-4} min^{-1}
Solar constant $= 2.0$ cal cm^{-2} min^{-1}

SATURATION MIXING RATIO (gm kg^{-1}) OVER WATER AT 1000 MB

					Temperature					
	0	*1*	*2*	*3*	*4*	*5*	*6*	*7*	*8*	*9*
40	49.8									
30	27.7	29.4	31.2	33.1	35.1	32.3	39.5	41.9	44.4	47.0
20	15.0	15.9	17.0	18.1	19.2	20.4	21.7	23.1	24.6	26.1
10	7.76	8.31	8.88	9.49	10.14	10.83	11.56	12.34	13.16	14.03
0	3.84	4.13	4.44	4.77	5.12	5.50	5.89	6.32	6.77	7.25
−0	3.84	3.57	3.31	3.08	2.85	2.64	2.45	2.27	2.10	1.94
−10	1.794	1.656	1.529	1.410	1.300	1.197	1.110	1.013	0.931	0.855
−20	0.785	0.720	0.659	0.604	0.552	0.505	0.461	0.421	0.384	0.350
−30	0.318	0.289	0.263	0.239	0.217	0.196	0.178	0.161	0.145	0.131

THE CORIOLIS FORCE

In Section 7.5 we discussed the effect of the rotation of the earth and suggested that this effect could be accounted for by introducing a force $2\Omega v \sin \phi$ acting on every unit mass moving with a velocity v. The force acts perpendicular to the direction of motion—to the right in the Northern Hemisphere and to the left in the Southern Hemisphere. The following does not give a rigorous proof of this value of the Coriolis force, but does suggest an explanation.

Consider a particle of unit mass at P (Figure A.1) on the earth's surface at latitude ϕ, moving northward with a velocity v. Let ϕ be measured in radians. The radius of rotation of the particle PM is $r = R \cos \phi$, where R is the radius of the earth. In time Δt, P moves northward a distance $v \Delta t$ to a latitude $\phi + \Delta \phi$ where

$$\Delta \phi = \frac{v \Delta t}{R}$$

If Ω is the angular velocity of the earth's rotation, the particle moves eastward $\Omega R \cos \phi \, \Delta t$.

The radius of rotation at the end of the path $= R \cos (\phi + \Delta \phi)$
$$= R(\cos \phi \cos \Delta \phi - \sin \phi \sin \Delta \phi)$$

Because $\Delta \phi$ is small, and measured in radians

$$\cos \Delta \phi \simeq 1 \qquad \text{and} \qquad \sin \Delta \phi \simeq \Delta \phi$$

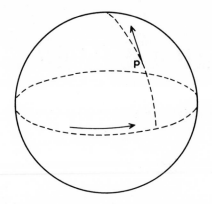

FIGURE A1 Motion of a particle northward on a rotating earth.

Hence, the radius of rotation is $r \simeq R \; (\cos \, \phi - \Delta\phi \, \sin \, \phi)$. At this latitude a particle fixed on the earth goes, in time Δt, a distance

$$\Omega R(\cos \, \phi - \Delta\phi \, \sin \, \phi) \, \Delta t$$

The moving particle gets ahead of the point fixed to the earth by a distance

$$
\begin{aligned}
s &= \Omega R \cos \, \phi \, \Delta t - \Omega R(\cos \, \phi - \Delta\phi \, \sin \, \phi) \, \Delta t \\
&= \Omega R \, \Delta\phi \, \sin \, \phi \, \Delta t
\end{aligned}
$$

Because $R \, \Delta\phi = v \, \Delta t$,

$$s = \Omega v \, \sin \, \phi(\Delta t)^2$$

This distance can be assumed to develop because of an imaginary acceleration a, given by

$$s = \frac{1}{2} \, at^2$$

which gives

$$\Omega v(\sin \, \phi)(\Delta t)^2 = \frac{1}{2} \, a(\Delta t)^2$$

$$a = 2\Omega v \, \sin \, \phi \tag{1}$$

For a unit mass, force equals acceleration. Therefore, the deviation from a northward movement is "explained" by an imaginary force equal to $2\Omega v \, \sin \, \phi$.

Now consider the particle moving eastward at latitude ϕ with a velocity v (Figure A.2). Again, the radius of rotation is $R \cos \, \phi$. A particle fixed to the earth under the moving particle is moving eastward at a velocity of $\Omega R \cos \phi$, while at the same time the moving particle is moving eastward at a velocity of $v + \Omega R \cos \phi$. By the law of centrifugal forces, both the moving and the fixed particles are subject to forces, perpendicular to the axis of the earth, whose magnitudes are given by the square of the velocity divided by the radius.

For the stationary particle, this force F_s is

$$F_s = \frac{\Omega^2 R^2 \cos^2 \phi}{R \cos \phi}$$

$$= \Omega^2 R \cos \phi$$

For the moving particle, the force F_m is

$$F_m = \frac{(v + \Omega R \cos \phi)^2}{R \cos \phi}$$

$$= \frac{v^2}{R \cos \phi} + 2\Omega v + \Omega^2 R \cos \phi \tag{2}$$

Each of these forces has two components—one perpendicular to the earth's surface, and one along the earth's surface acting to the south. For F_s the component perpendicular to the earth's surface is $F_s \cos \phi$. This acts against the gravitational force and, therefore, causes the particle to weigh less than it would on a stationary earth. The component along the earth's surface is $F_s \sin \phi$. Every particle is acted on by

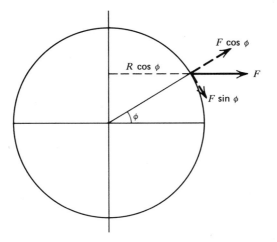

FIGURE A2 Forces acting on a particle moving eastward on a rotating earth.

a force tending to move it toward the equator. The opposing force arises because the earth is not a perfect sphere, and a southward movement would take the particle farther from the center of the earth. The shape of the earth is such that these two opposing forces balance, and the net force is zero.

For the moving particle the last term of Equation 2 is thus cared for by the shape of the earth. The difference $F_m - F_s$ arises because of the motion of the particle

$$F_m - F_s = \frac{v^2}{R \cos \phi} + 2\Omega v$$

Again, the component perpendicular to the surface $(F_m - F_s) \cos \phi$ causes the moving body to weigh less, but otherwise has no apparent effect. The other component $(F_m - F_s) \sin \phi$ tends to make the body move southward. The value of $v^2 / R \cos \phi$ is generally small compared with $2\Omega v$ and may be ignored. Thus a force $2\Omega v \sin \phi$ causes the body to move southward unless another force prevents it.

Thus we observe that with a body moving either along longitude lines or along latitude lines, a force $2\Omega v \sin \phi$ tends to deflect a moving mass toward the right of its path. This is the Coriolis force. A similar discussion would show that south of the equator this force must act to the left of the direction of motion.

HOURLY WEATHER REPORTS FOR 1200 h GMT 8 JANUARY TO 1200 h GMT 9 JANUARY 1969 FOR VARIOUS STATIONS

The following give the hourly weather reports for a number of stations for the period from 1200 GMT 8 January until 1200 GMT 9 January 1969. This is the period studied in Chapter 10. The reports should be studied along with the weather maps, Figures 10.2, 10.3, and 10.4. By means of such sequences of reports, one discovers the variety of weather changes as a front passes a station.

These reports are in code in order to put as much information as possible into a brief report. The first group gives the day and hour, Greenwich Mean Time, of the report. The second group gives information on the layers of cloud above the station, the visibility and the current weather. A cloud layer may be scattered, ⓓ, broken, ⓪, or overcast, ⊕. A thin layer is shown by a – sign preceding the symbol. When the sky is obscured, as by fog or snow, the symbol X is used. Clear skies are shown by ○.

The height of the cloud is given, in hundreds of feet, preceding the cloud symbol. This value may have been obtained by estimation, E, or measurement, M. An indefinite cloud base is indicated by W. High clouds are preceded by a slant.

The number following the last cloud symbol gives the visibility in miles. If this is not uniform in all directions, this is noted by the letter V. In reporting weather at the end of the cloud and weather group, rain, snow, snow grains, fog, ground fog, dust, haze, blowing snow, and blowing dust are coded by their initial letters. Drizzle is L, and smoke is K. Z added to R or L indicates that the precipitation is freezing on impact. RW and SW indicate showery weather. The symbol – indicates that the weather phenomenon is light.

The last group gives the sea-level pressure, temperature (°F), dew point (°F), and wind velocity, separated by slants. The first two figures of the wind group give the direction in tens of degrees from north through east. The second two figures gives the mean speed in miles per hour. With gusty winds, the top gust is given at the end separated from the mean wind by the letter G.

Columbia, Missouri (38°58′N 92°22′W)

0812	/-ⓓ15	966/29/16/1614G19
0813	100ⓓ/⊕15	951/30/19/1611G20
0814	100ⓓ/⊕15	931/31/19/1514
0815	100ⓓ/⊕15	926/33/21/1514
0816	/⊕15	939/37/24/1411
0817	/⊕15	927/41/27/1512
0818	M26⊕15	914/44/30/1410
0819	M25⊕15	910/45/33/1806
0820	M23⊕15	920/46/34/2806
0821	M15⊕12	938/41/34/3020
0822	M12⊕12	969/35/31/3123G28
0823	M17⊕12	989/34/27/3217G23
0900	M19⊕10	017/34/28/3118G24
0901	M22⊕10	038/33/25/3114G20

Columbia, Missouri (38°58′N 92°22′W) (Continued)

0902	22⊙10	048/31/19/3122G28
0903	22⊙10	061/29/16/3117G28
0904	○10	076/27/18/3120G27
0905	○10	092/23/11/3118G25
0906	○12	116/19/7/3119G24
0907	○15	126/17/8/3116G24
0908	○15	136/15/3/3215G24
0909	○15	151/13/3/3116G23
0910	○15	159/12/0/3114G20
0911	○15	166/11/0/3113G17
0912	○15	176/10/1/3110

Springfield, Missouri (37°14′N 93°23′W)

0812	/⊙15	956/31/21/1615G21
0813	/–⊙15	941/33/24/1714G22
0814	/–◐15	927/35/26/1620G27
0815	150⊙/⊕10	910/38/30/1617G27
0816	/⊕12	902/45/34/1718G26
0817	E150◐/⊕12	905/47/37/1815G22
0818	150⊙/⊕12	908/52/38/2103
0819	150⊙/⊕12	904/56/39/2512
0820	20⊙/⊕15	920/59/38/3014G22
0821	M9⊕4H	959/42/35/3216G27
0822	M12⊕5H	984/40/32/3214G20
0823	M13⊕5H	014/37/29/3218G27
0900	M17⊕6H	041/36/27/3215G24
0901	34⊙M150⊕10	064/34/24/3317G24
0902	E150⊕12	080/34/20/3315G22
0903	150⊙/◐15	093/33/18/3315G21
0904	E150◐15	106/31/10/3215G21
0905	/⊙15	110/29/6/3217G23
0906	150⊙/–◐15	114/28/7/3214G20
0907	/◐15	119/28/6/3313G19
0908	150⊙/◐15	131/25/7/3315G22
0909	150⊙/◐15	151/23/6/3315G21
0910	○15	158/20/7/3412G18
0911	150⊙/⊙15	173/17/5/3315G20
0912	100⊙15	181/15/5/3413

Fort Smith, Arkansas (35°20′N 94°22′W)

0812	○20	976/34/27/0712
0813	/–◐20	969/36/28/0910
0814	/–◐20	958/35/29/0406
0815	/–◐20	948/40/32/0708
0816	/–◐20	940/48/37/0708
0817	/–◐20	926/66/47/2314G22
0818	/–◑20	919/69/45/2314G22
0819	/–◑20	911/73/44/2514G23
0820	/–◑20	911/75/42/2715G23
0821	/–⊕20	921/75/35/2918G26
0822	/–◑20	934/71/32/3017G24
0823	/◑15	987/55/36/3117G23
0900	/◑10	029/47/33/3513G19
0901	M23⊕8	055/43/30/3517G28
0902	M29⊕8	081/43/29/0116G24
0903	30◐/◑8	100/40/25/0214G19
0904	/◑10	115/39/24/0115G22
0905	/◐10	125/38/17/0114G19
0906	○12	137/37/18/3615
0907	○20	148/36/6/3613G20
0908	○20	150/36/5/3512G18
0909	30◐20	155/35/6/3210
0910	30◐/◑20	160/35/7/3413
0911	/⊕20	174/35/7/3415
0912	E150◑20	179/33/8/3511

Memphis, Tennessee (35°03′N 89°59′W)

0812	M14◑45⊕12	046/48/38/1708
0813	M14V⊕12	046/49/39/1707
0814	M12⊕8	036/51/42/1611
0815	M10⊕6K	025/53/45/1707
0816	M14⊕6H	014/56/49/1715G22
0817	M17⊕6H	992/60/52/1718G26
0818	M22⊕5BD	976/65/54/1820G32
0819	M25◑5BD	953/68/55/2020G32
0820	E25◑5BD	939/69/55/2032G42
0821	E30◑5BD	935/70/55/2020G30
0822	30◐/◐8	945/70/55/2230G45
0823	30◐/–◑8	946/69/55/1920G30
0900	40◐/–◑10	956/68/55/1918G28

Memphis, Tennessee (35°03′N 89°59′W) (Continued)

0901	E60⫯/⊕10	970/68/55/2116
0902	E65⫯10	975/68/55/2214
0903	E65⫯10	978/68/56/2015G23
0904	M33⊕8R−	018/57/45/3017G30
0905	M40⊕10	044/52/39/3214
0906	M45⫯10	061/47/33/3214G20
0907	M60⫯15+	080/45/31/3212
0908	M70⫯15+	094/44/30/3210
0909	M80⫯15+	106/43/24/3209
0910	M75⊕15+	111/42/20/3111
0911	75⊙15+	121/38/17/3409
0912	○15+	130/35/16/3508

Nashville, Tennessee (36°07′N 86°41′W)

0812	E50⊕12	094/31/14/1409
0813	E50⊕12	094/32/15/1308
0814	M25V⊕8R−−	083/34/17/1409
0815	M25V⊕8R−−	079/39/23/1612
0816	M20⫯50⊕8	069/42/25/1713G20
0817	M20⫯50⊕10	056/47/29/1812G20
0818	M20⊕10	035/49/33/1715G22
0819	M22⊕8	015/49/35/1713G21
0820	23⊙E45⫯/⫯8	993/53/39/1717G24
0821	20⊙E45⫯/⫯8	983/52/42/1820
0822	E25⫯50⊕8	965/54/45/1817G24
0823	M27⊕8	975/55/47/1814G22
0900	M28⊕10	984/58/48/1915G22
0901	M30⊕10	989/58/47/1713
0902	M35⫯/⊕10	985/58/48/1815
0903	E60⫯/⊕12	981/61/50/1918G27
0904	E60⫯/⊕12	974/62/50/1919G27
0905	E60⫯/⫯12	974/63/51/2018G25
0906	M30⊕8RW−	983/63/52/2118G28
0907	7⊙M18⊕6R−	016/51/48/3214G23
0908	7⊙M13V⊕6R−	037/48/44/3317G22
0909	M6⊕7	056/44/40/3217
0910	M12⊕7	077/40/35/3215
0911	M14⊕7	087/37/31/3312
0912	M16⫯23⊕7	100/35/29/3414

Madison, Wisconsin (43°08′N 89°20′W)

0812	M70⊕12S−−	055/14/9/1217
0813	M65⊕12S−−	038/15/9/1315G20
0814	M60⊕15	024/15/9/1418
0815	M40⊕15	009/17/8/1416G22
0816	M40⊕15	015/18/9/1413
0817	−XE40⊕1½S−	000/20/13/1315G20
0818	W7X1S−F	985/20/16/1416G19
0819	M34⊕4S−−GF	986/20/15/1609
0820	M31⊕7	958/20/16/1313
0821	M30⊕7	958/22/15/1410
0822	M25⊕2½S−F	969/22/16/1206
0823	W13X1VS−F	960/22/18/1205
0900	M25⊕2S−−F	956/22/18/1406
0901	M33⊕3S−−F	950/22/18/2007
0902	W11X2VSG−F	950/22/18/2207
0903	M14⊕7	949/22/18/2310
0904	M13◍33⊕6S−−	952/22/18/2714G22
0905	M23◍12	962/19/11/2714G22
0906	M20◍12	971/15/8/2718G30
0907	M20◍12	986/9/3/2722G30
0908	20◉15	012/5/−4/2718
0909	◯15	020/3/−5/2715G23
0910	◯15	023/1/−7/2718
0911	◯15	026/1/−8/2618G25
0912	◯15	038/0/−8/2620G30

Springfield, Illinois (39°50′N 89°40′W)

0812	M44⊕15	032/23/15/1316
0813	M36⊕15+	019/24/16/1314
0814	M39⊕15+	999/25/16/1216
0815	M34⊕8	998/26/17/1215
0816	M17⊕7	985/29/19/1313
0817	M16⊕7	968/31/21/1212
0818	M18⊕6H	950/32/23/1314G18
0819	M16⊕6H	929/33/24/1413
0820	M12⊕6H	917/33/26/1513
0821	M10⊕6H	927/34/28/1610
0822	M8⊕2½L−F	932/34/30/2304
0823	M4X3/4L−F	950/32/30/2704
0900	M4◍8◍20⊕3L−F	956/32/30/2809

Springfield, Illinois (*39°50′N 89°40′W*) (Continued)

0901	4①M10⊕8ZL--	981/31/29/2915G23
0902	8①M16①28⊕10	995/31/26/2913G22
0903	M17①30⊕10	001/31/25/3014G23
0904	M16⊕10	023/29/22/3023G36
0905	M24⊕10	049/23/15/3023G33
0906	30①100①15+	070/17/8/3020G32
0907	/-①15+	088/14/5/2821G27
0908	/-①15+	098/13/4/2920G27
0909	○15+	105/12/3/2715G24
0910	○15+	119/11/2/2716G23
0911	○15+	123/10/1/2820G25
0912	○15+	135/9/1/2717G24

Grand Rapids, Michigan (*42°54′N 85°40′W*)

0812	M75⊕15	113/11/8/1406
0813	M40⊕12	113/13/9/1506
0814	M23⊕9	112/14/9/1210
0815	W7X1½VS-	104/15/10/1310
0816	W7X1½VS-	090/15/11/1214
0817	30①E50⊕12S-	066/16/12/1118
0818	15①E65⊕12	066/17/12/1212G21
0819	15①M50⊕15	042/17/12/1215G23
0820	M18⊕4S-	032/17/12/1115G22
0821	M25⊕6S-	011/17/12/1116G22
0822	M40⊕8S-	999/17/13/1016G22
0823	M26⊕8	992/17/13/1018G24
0900	W10X2S-	984/17/13/0918G24
0901	W8X1VS-	977/17/14/0917G22
0902	W6X1VS-	968/18/15/0914G18
0903	M22⊕2S-	956/18/16/1014
0904	M6⊕2S-F	936/19/17/1013G17
0905	M6⊕3S-F	945/19/17/1307
0906	M5⊕2S-F	941/20/17/1003
0907	M18⊕2VS-F	938/20/17/1203
0908	M6⊕10	944/20/18/2003
0909	M13⊕8S--	950/22/21/2713
0910	M16⊕6S-	954/24/19/2717G23
0911	11①M16⊕3S-BS	967/20/17/2621G32
0912	M13⊕2½S-BS	977/19/17/2618G28

Cleveland, Ohio (41°24′N 81°51′W)

0812	10①/-⨁6H	162/3/-4/1907
0813	/-⨁6H	155/4/-3/1908
0814	/-⊕6H	151/6/1/1907
0815	/-⨁6H	155/9/2/1905
0816	150①/-⊕6H	155/12/5/1907
0817	150①/⊕6H	136/16/7/1608
0818	100①/⊕6H	109/18/7/1507
0819	E100⨁/⊕6H	109/19/8/1510
0820	E100⊕6H	097/20/9/1610
0821	W12X3/4S-H	092/20/11/1312
0822	M23⨁50⊕5S--H	080/20/12/1212
0823	M44⊕7	060/21/12/1313
0900	M42⊕6S--	050/21/12/1315
0901	M35⊕5S--	026/22/12/1218
0902	M34⊕7	012/23/12/1317
0903	M24⊕5ZL--H	992/25/13/1510
0904	M16⊕5H	970/27/20/1612
0905	M12⊕5ZR-H	962/28/22/1712
0906	M10⊕5H	946/29/25/1714
0907	M8⊕5H	938/32/28/1816
0908	M6⊕2L--F	949/33/29/2015
0909	M4⊕1½R-F	954/33/30/2412
0910	M7⊕4R--F	963/34/29/2715
0911	M8⊕7	974/33/26/2613G21
0912	E10⊕7	989/32/24/2817

Columbus, Ohio (40°00′N 82°53′W)

0812	E90⊕12	166/7/1/1206
0813	E75⊕7	158/9/3/1305
0814	80①/⊕5HK	157/11/4/1807
0815	80①/⊕6HK	149/14/5/1207
0816	E80⨁150⊕6HK	141/16/6/1311
0817	E80⨁150⊕7	126/17/7/1212
0818	E80⊕7	103/19/9/1211
0819	E80⊕7	076/20/9/1114
0820	M36⨁80⊕7	076/21/9/1212
0821	M32⊕7	061/22/10/1112
0822	M23⊕7	050/22/11/1112
0823	M21⊕7	035/23/11/1214
0900	M21⊕7ZR-	020/23/11/1217

Columbus, Ohio (40°00′N 82°53′W) (Continued)

0901	M17⊕6ZL–	007/24/17/1412
0902	M13⊕5ZR–H	997/25/19/1211
0903	M16⊕6ZR–H	975/28/24/1410
0904	M17⊕5ZR–H	975/30/25/1410
0905	M12⊕4ZL–F	967/31/27/1707
0906	M9⊕5ZL– –F	962/32/28/1808
0907	M6⊕5F	954/33/30/1710
0908	M4⊕4F	971/35/31/2312
0909	6①M12⊕7	990/36/31/2814G24
0910	M10⊕7	002/35/29/2710G20
0911	M14⊕10	018/32/25/2715G25
0912	M19⊕10	035/30/20/2712G20

BIBLIOGRAPHY

This bibliography gives a list of books and magazines for the nonprofessional, although some of the books listed have values for the professional meteorologist as well.

Magazines

The following magazines are designed for the nonprofessional. They contain clear descriptions of meteorological phenomena, and some discussion of their causes. The last two have also excellent pictures as illustrations.

Meteorological Magazine, published by the Meteorological Office, Great Britain, has many popular articles on meteorology.
Weather, published by the Royal Meteorological Society, 49 Cromwell Road, London SW 7, England.
Weatherwise, published by the American Meteorological Society, 45 Beacon Street, Boston, Massachusetts 02108.

Books on Elementary Meteorology

Barry, R. G., and R. J. Chorley, *Atmosphere, Weather, and Climate.* London: Methuen. 1968, 311 pp.

Battan, Louis J., *The Nature of Violent Storms.* New York: Anchor Books— Doubleday and Co. 1961, 158 pp.

Blair, Thomas A., and Robert C. Fite, *Weather Elements: A Text in Elementary Meteorology.* Englewood Cliffs, N.J.: Prentice-Hall, 1964. 5th ed. 364 pp.

Dobson, G. M. B., *Exploring the Atmosphere.* Oxford: Oxford University Press. 1963. 188 pp.

Donn, William L., *Meteorology.* New York: McGraw-Hill, 1965. 3rd ed. 484 pp.

Great Britain—Meteorological Office, *A Course in Elementary Meteorology.* London, Her Majesty's Stationery Office. 1960. 189 pp.

Great Britain—Meteorological Office, *Handbook of Aviation Meteorology.* London, Her Majesty's Stationery Office. 1960. 390 pp.

Hare, F. K., *The Restless Atmosphere*. New York: Harper and Row. 1963. 192 pp.

Miller, A., *Meteorology*. Columbus, Ohio: Charles E. Merrill Books. 1966. 128 pp.

Neuberger, H., and F. B. Stephens, *Weather and Man*. Englewood Cliffs, N.J.: Prentice-Hall. 1948. 272 pp.

Petterssen, S., *Introduction to Meteorology*, New York: McGraw-Hill, 1969. 3rd ed. 333 pp.

Riehl, Herbert, *Introduction to the Atmosphere*. New York: McGraw-Hill. 1965. 365 pp.

Riehl, Herbert, *Tropical Meteorology*. New York: McGraw-Hill. 1954. 392 pp.

Stewart, G. R., *Storm*. New York: Modern Library. 1947. 349 pp. This book is written as fiction, but the account tells much about the nature of a storm and its effects.

Sutcliffe, R. C., *Weather and Climate*. New York: W. W. Norton & Co. 1966, 206 pp.

Sutton, O. G., *The Challenge of the Atmosphere*. New York: Harper and Brothers. 1961. 227 pp.

Thompson, Philip D., and Robert O'Brien, *Weather*. New York: Time Inc. 1965. 200 pp., with numerous photos.

Van Straten, Florence W., *Weather or Not*. New York: Dodd, Mead, & Co. 1966. 237 pp.

Wallington, C. E., *Meteorology for Glider Pilots*. London: J. Murray 1961. 284 pp.

Books on Climatology

Brooks, C. E. P., *Climate in Everyday Life*. New York: Philosophical Society Library 1951. 314 pp.

Critchfield, H. J., *General Climatology*. Englewood Cliffs, N.J.: Prentice-Hall, 1966. 2nd ed. 420 pp.

Kendrew, W. G., *The Climates of the Continents*. Oxford: Oxford University Press, 1961. 5th ed. 608 pp. An excellent reference book.

Lamb, H. H., *The Changing Climate*. London: Methuen. 1966. 236 pp.

Landsberg, H., *Physical Climatology*. Dubois, Pa.: Gray Printing Co., 1958. 2nd ed. 446 pp.

Manley, Gordon, *Climate and the British Scene*. London: Collins. 1952. 314 pp.

Miller, A. A., *Climatology*. London: Methuen, 1961, 9th ed. 320 pp.

Rummey, George R., *Climatology and the World's Climates*. New York: Mac-Millan 1968. 656 pp.

Sellers, William D., *Physical Climatology*. Chicago: University of Chicago 1965. 272 pp.

Shapley, Harlow (ed.), *Climatic Change: Evidence, Causes, and Effects.* Cambridge, Mass.: Harvard University Press. 1954. 318 pp.

Trewartha, Glenn T., *An Introduction to Climate.* New York: McGraw-Hill, 1968. 4th ed. 408 pp.

U.S. Weather Bureau, *Decennial Census of the United States Climate.* Washington D.C. 1963. 318 pp., entirely tables.

Books on Special Topics Dealing with Meteorology

Barrett, E. C., *Viewing Weather from Space.* London: Longmans. 1967. 140 pp.

Battan, Louis J., *Cloud Physics and Cloud Seeding.* New York: Anchor Books—Doubleday & Co. 1962. 144 pp.

Battan, Louis J., *Unclean Sky: a Meteorologist Looks at Air Pollution.* Garden City, N.Y.: Doubleday & Co. 1966. 141 pp.

Battan, Louis J., *The Thunderstorm.* New York: Signet Science Library. 1964. 128 pp.

Bell, Corydon, *Wonder of Snow.* New York: Hill and Wang. 1957.

Bruce, J. P., and R. H. Clark, *Introduction to Hydrometeorology.* London: Pergamon Press. 1966. 319 pp.

Dunn, Gordon, and Banner I. Miller, *Atlantic Hurricanes.* Baton Rouge: Louisiana State University Press, 1964. 2nd ed. 377 pp.

Flora, S. D., *Hailstorms of the United States.* Norman, Okla.: University of Oklahoma Press, 1956, 2nd ed. 201 pp.

Flora, S. D., *Tornadoes of the United States.* Norman, Okla.: University of Oklahoma Press. 1954. 221 pp.

Geiger, Rudolf, *The Climate near the Ground.* Cambridge, Mass.: Harvard University Press, 1965. 4th ed. 611 pp. Filled with detail of the variations of climate in short distances.

Herber, Lewis, *Crisis in Our Cities.* Englewood Cliffs, N.J.: Prentice-Hall. 1965. 239 pp.

Lane, Frank W., *The Elements Rage.* Philadelphia: Chilton Books. 1965. 346 pp. Written in a popular style, and giving vivid details about storms and their effects.

Lewis, Alfred, *Clean the Air!* New York: McGraw-Hill. 1965. 96 pp.

Licht, Sidney, and H. L. Kamenetz (eds.), *Medical Climatology.* New Haven, Conn.: Elizabeth Licht. 1964. 753 pp.

Lowry, William P., *Weather and Life.* Corvallis, Oregon: O.S.U. Book Stores. 1967. 220 pp.

Ludlum, David M., *Early American Hurricanes 1492–1870.* Boston: American Meteorological Society. 1963. 200 pp.

Ludlum, David M., *Early American Winters 1604–1870.* Boston: American Meteorological Society. 1966. 285 pp..

Ludlum, F. H., and R. S. Scorer, *Cloud Study—A Pictorial Guide*. London: Royal Meteorological Society. 1958. 80 pp.

Malan, D. J., *Physics of Lightning*. London: English University Press. 1963. 176 pp.

Mason, Basil J., *Clouds, Rain and Rainmaking*. Cambridge, England: Cambridge University Press. 1962. 145 pp.

McClement, Fred, *The Anvil of the Gods*. Philadelphia: J. B. Lippincott. 1964. 272 pp. Treats the thunderstorm and its effect on aircraft.

Middleton, W. E. K., *History of the Thermometer*. Baltimore, Md.: John Hopkins Press. 1966. 249 pp.

Middleton, W. E. K., *A History of the Theories of Rain*. New York: Franklin Watts Inc. 1965. 223 pp.

Middleton, W. E. K., *Vision through the Atmosphere*. Toronto: Toronto University Press. 1952. 250 pp.

Ross, Frank X., Jr., *Weather: the Science of Meteorology from Ancient Times to the Space Age*. New York: Lothrop, Lee and Shepard. 1965. 200 pp.

Schonland, Basil, *The Flight of Thunderbolts*. Oxford: The Clarendon Press, 1964. 2nd ed. 182 pp.

Scorer, Richard S., and Harry Wexler, *Colour Guide to Clouds*. Oxford: Pergamon Press. 1963. 63 pp.

Tannehill, Ivan R., *Drought, its Causes and Effects*. Princeton, N.J.: Princeton University Press. 1947. 264 pp.

Viemeister, P. E., *The Lightning Book*. New York: Doubleday & Co. 1961. 316 pp.

Wagler, David, *The Mighty Whirlwind*. Aylmer, Ontario: Pathway Pub. Corp. 1966. 226 pp.

ANSWERS TO PROBLEMS

CHAPTER 1   3. 17°C

4. The storm releases energy equivalent, approximately, to 500 megaton bombs. The energy release of the storm is over a larger volume and over a longer period of time, and so does not cause the damage that would result from the explosion of the bombs.

CHAPTER 2   4. 1.4 m

7. 1.5 cal hr^{-1} 9. 5°C

8. 34 cal cm^{-2} 10. 2.5°C

CHAPTER 3   1. 75×10^7 kg

2. 1250 mb. 38,600 cal

4. 2.7×10^5 kg. 1.1×10^2 kg

5. By using a mean pressure of 400 mb, the difference is 3440 m. The correct value is 3518 m. The error arises because in reality the mean pressure cannot be taken as the arithmetic mean between the lower and upper pressures. The error becomes magnified for thick layers. The temperature difference calculated by using 400 mb is 3.4°C, which is approximately correct.

6. Sea-level pressure below A: 1019 mb by using a mean pressure of 900 mb and a mean temperature of 7°C, with the more accurate formula, 1021 mb. Corresponding values for B are, respectively, 1028 and 1030 mb. Pressure difference is 9 mb.

7. 1018 mb

8. 778 mb (780 mb by using Equation 6)

9. 44 mb

10. Approximately, 0.1 in. change represents a height change of 100 ft. The value varies from 75 ft at −20°C to 84 ft at 30°C.

11. 1019.2 mb; 1018.8 mb

13. 0.9 mb

CHAPTER 4   4. 253 gm

5. 9.9 kg

6. 6°C; −9°C

7. 1.2 per cent

12. 0.17 cm

13. Relative humidity = 50 per cent

CHAPTER 5   1. 43 per cent

5. 7 per cent

CHAPTER 6   1.

	T (°C)	T_D (°C)	w (gm kg^{-1})	T_w (°C)	LCL (mb)	RH (per cent)	θ (°C)	θ_w (°C)
a	15.7	12.2	9.0	13	960	80	16	13
b	6.3	−4	3.3	1.0	790	50	13	5
c	−1	−6	3.0	−3	740	67	17	8
d	−25	−28	0.4	−25	985	75	−27	−27
e	25	10	8.0	15	765	40	29	17

2. 7.3 gm kg^{-1}; 1.1×10^6

3. 48°C

4. (a) (1) 206 mb; (2), (3), (4), nil; (6) layer 914–760 shows slight instability.

(b) (1) 169 mb; (2), (3), nil; (4) 1002 mb, 850 mb (slight); (5) 940 mb, 160 mb.

(c) (1) 174 mb; (2) nil; (3) 850–670 mb; (4) nil.

(d) (1) 326 mb; (2) 1024–904 mb; (3), (4) nil.

(e) (1) 260 mb; (2) 995–981 mb; 927–877 mb; (3) 1015–995 mb; (4) nil.

5. $8\frac{1}{2}$°C

6. −20.5°C. 3°C

7. Final conditions: $T = 0$°C; $w_s = 5.5$ gm kg^{-1}

9. 39°C

11. 230 gm of air hr^{-1}; 1.2 m sec^{-1}

CHAPTER 7   1. 353 km; 3.53 cm

For v = 40 30 20 15 10   5 m sec^{-1}

Spacing of isobars = 0.9 1.2 1.8 2.4 3.5 7.1 cm

2. 16 m sec^{-1}

4. $V_{77°}/V_{-40°} = 1.28$

5. 6.3×10^{-4}

6. The wind blows along AB; working with three significant figures, the difference is 70 m accurate to one figure only; more accurate calculations give 69 m; 34 m sec^{-1}.

7. 8.1 km

8. By approximate methods, 2560 m; by more accurate methods, 2593 m; 10°C

INDEX